The Dynamic Neuron

The Dynamic Neuron

A Comprehensive Survey of the Neurochemical Basis of Synaptic Plasticity

John Smythies

A Bradford Book
The MIT Press
Cambridge, Massachusetts
London, England

This book was set in Palatino by Achorn Graphic Services, Inc., and was printed and bound in the United States of America.

Library of Congress Cataloging-in-Publication Data

Smythies, John R. (John Raymond), 1922–
  The dynamic neuron : a comprehensive survey of the neurochemical basis of synaptic plasticity / John Smythies.
    p. cm.
  "A Bradford book."
  Includes bibliographical references and index.
  ISBN 0-262-19473-2 (hc. : alk. paper)
  1. Neuroplasticity. 2. Neurochemistry. I. Title.

QP363.3 .S525   2002
573.8′36—dc21

                                                              2001058700

Contents

Preface

My interest in synaptic plasticity arose out of my researches in schizophrenia. In 1952 Humphrey Osmond and I published the first specific biochemical hypothesis of schizophrenia. I have been active in this field ever since. The reports that schizophrenics have lost some 50% of their dendritic spines in the cortex led me to an intensive investigation into the factors that control spine numbers and synaptic plasticity. Only by a thorough understanding of normal function can one hope to achieve an understanding of the causes of disease in this system. These factors turned out to be exceedingly complex and led to fields as diverse as the detailed mechanisms operative at synapses, in particular the glutamate synapse; redox factors such as oxidative stress and antioxidant protection; a great variety of signaling cascades; and the recent advances in cell biology concerning cell adhesion molecules, endocytosis, scaffolding proteins, and neurotropins.

Current scientific endeavors are churning out a vast amount of data which represent innumerable small pieces of a giant jigsaw puzzle. However, little attention is being paid to fitting this vast mass of individual facts into a single coherent account of what is going on. As Christof Koch (personal communication, 2000) has said, "We are drowning in a sea of data." The aim of this book is to try to remedy this situation in one small branch of neuroscience—synaptic plasticity. The book attempts to cover the significant parts of two normally widely separate disciplines—neuroscience and cell biology. As a neuroscientist, therefore, I am happy to have had the help of a leading cell biologist, Renate de Wit.

The book is written for a general scientific audience with an interest in synaptic plasticity. Therefore some of the neuroscience part is targeted to cell biologists and some of the cell biology part is targeted to neuroscientists. Neuroscience and cell biology today are inseparably linked, and both need to be taken into account by all scientists interested in neurons.

A word on the style and format of the book may be helpful to the reader. Most books of this kind are divided into chapters written by a different author or group of authors. These are each experts in their own particular field and present a selective and critical discussion of the data in their field, rather than casting a wide net and including all the existing data. The drawback of this method is that there is usually little correlation between the different chapters. However, this option is not open to the single author who tries to cover an entire field. In my case, I am not an expert in most of the fields I discuss in this book. My professional expertise is limited to redox systems and to the neurochemistry of schizophrenia. Therefore the book's discussion must include critical discussions already made by experts in the various subfields covered, which I have done. The benefit of this method is that hitherto unsuspected but important connections may be found between material in the different subfields that could only have been discovered by an inclusive, as opposed to a selective, approach. I have listed several such advances in this book.

After a general introduction in chapter 1, section 1.2 covers the evidence that there is in fact synaptic plasticity in the brain. Sections 1.3 to 1.6 then examine at length the important features of the glutamate synapse that actually engineer this synaptic plasticity, with particular emphasis on my own field of redox systems. Chapter 2 then discusses the key roles of membrane endocytosis and exocytosis in neuronal plasticity. Chapter 3 discusses various cell biology mechanisms, such as cell adhesion molecules, scaffolding proteins, axon-directing proteins, neurotropins, actin, and local protein synthesis. Chapter 4 deals with miscellaneous topics. Chapter 5 covers pharmacological and clinical implications of these new developments.

A list of acronyms and abbreviations is given at the end of the book.

Acknowledgments

I am grateful to Francis Crick, who first stimulated my interest in what makes spines twitch, and to Vilanayur Ramachandran and Michael Trimble for their continued support of this work. Also very helpful have been comments and communications from a number of people in the field, including Eric Altschuler, Ashley Bush, Joseph Coyle, Jenny Koenig, Richard Kostrewza, Jiankang Liu, Peter O'Brien, Jack Pettigrew, Juan Segura-Aguilar, Terrence Sejnowski, and Richard Wyatt. Renate de Wit very kindly acted as a consultant on cell biology.

Chapter 1

Synaptic Biochemistry

1.1 Introduction

Until quite recently, the picture of the synapse and its connections presented by neuroscience was based on two familiar notions. The first was the computer with its permanent chips, its hardwired connections, and its binary code. The second was the familiar static photographs produced by the electron microscope. This led to the belief that individual synapses represented more or less permanent connections between neurons that operated a binary code firing (i.e., producing an axonal action potential) and not firing (no action potential). Thus the computations performed by the brain could be described exclusively as the results of fixed nerve nets operating by such processes as matrix multiplication (Churchland and Sejnowski 1992). Learning was supposed to depend largely on a change in weights at individual synapses, which altered the probability that activity at that synapse would contribute to firing its postsynaptic neuron.

In a similar manner, elements of the synapse (membrane, receptors, postsynaptic density, cytoskeleton, etc.) were also regarded as semipermanent structures subject to only a slow metabolic turnover. Neurotransmitter molecules released from axon terminals crossed the synaptic cleft and bound to their particular receptors. This initiated a conformational change in the receptor protein, which in turn either opened an ion channel or, in the case of G-protein-linked receptors, initiated a postsynaptic biochemical cascade mediated by second messengers (small molecules such as cyclic nucleotides and phosophoinositides) and resulted in protein phosphorylation.

The action of the neurotransmitter was terminated by its departure from the receptor molecule and its subsequent reuptake into the presynaptic terminal for reuse (e.g., glutamate, Glu) or its further metabolism to inactive products (e.g., acetylcholine, ACh, and dopamine). The receptor remained in the membrane, awaiting the arrival of another neurotransmitter molecule, whereupon the same process would be repeated. It was recognized that the receptor would eventually be replaced by a newly synthesized receptor molecule, but this was thought to be a slow process. Moreover, the receptor and other associated transmembrane proteins were thought to be freely mobile and to float in the membrane like icebergs in the sea, moving to make contact with other molecules, such as G-proteins, when necessary.

The interior of the postsynaptic neuron was thought to be a sort of soup in which biochemical reactions took place between (relatively) freely mobile molecules, much as they did in the test tubes where the neurochemists were used to observing them. The organelles (e.g., mitochondria, nucleus, endoplasmic reticulum, Golgi apparatus, and endosomes) floated in this biochemical soup. The focus in neurochemistry was largely on the biochemical reactions themselves. Little attention was paid to the structural organization of all these organelles and to the actual compartmentalization of these biochemical reactions; that is, where in the cellular machinery all these reactions take place (with the partial exception of the mitochondria).

Recent research, however, has shown that this classic picture is wildly inaccurate. Most synapses are not fixed structures, but are subject to a process of continual pruning, and replacement by new synapses. Learning depends on a change in synaptic weights at individual synapses to some degree, particularly in primary sensory and motor cortices. However, it also depends on the continual removal of old synapses and the growth of new ones (particularly in the association cortex and the limbic system). Receptors, and indeed the whole external membrane of the neuron, are subject to a continual dynamic process of rapid internalization into the postsynaptic neuron, where they are processed by the endosome system. Some are then recycled to the surface and reused. The rest are catabolized by proteosomes[1] and lysosomes.[2]

The interior of neurons is not a soup, but is highly structured, in part by a variety of scaffolding proteins, so that interacting enzymes and their substrates are held in highly specific microanatomical relation to each other. The system is full of marvelously engineered little protein machines, with precise and meticulous locations, acting in production lines that Henry Ford would have been proud of. The mechanism includes processing and sorting areas connected by "railway lines" powered by small protein shunting engines that deliver each product of the synthetic machinery to its correct destination (for a good review of these motor proteins, see Cross and Carter 2000). These molecules are often wrapped in specific delivery vesicles.

This whole system is controlled by a vastly complex "signaling" system. Membrane-bound receptors themselves do not float freely in the membrane, but are tied to scaffolding proteins. This locates and orients them precisely with respect to the other proteins in the postsynaptic cascades with which they interact (Schwencke et al. 1995). Lysakowski et al. (1999, p. 388) point out that "the . . . 'free' intracellular space is extremely limited." Tanaka et al. (2000, p. 388) state that the static electron micrograph picture of the synapse requires "extensive reevaluation." Lemmon and Traub (2000, p. 463) say that "membrane flow through the endosome system compartment is both enormously complex and precisely regulated."

Many chains of molecular events I describe in this book have been termed "signaling" cascades. Some of these, such as nitric oxide (NO) and hydrogen peroxide, really do carry signals. However, many of the protein–protein interactions (as well as other molecular interactions) to be described are not really signaling cascades, but instead are very complex machines. When I tread on the brake pedal of my car, the machinery between the brake pedal and the brake hubs does not carry any signal that says Stop!. It is simply a machine designed to connect activities.

Likewise, the myriad mainly protein–protein interactions inside cells cause biochemical event *A* (e.g., binding a transmitter molecule at its receptor) to bring about at a distance biochemical event *B* (e.g., modulation of DNA). It is somewhat anthropomorphic to describe this process in terms of signaling rather than in terms of molecular machinery marvelously engineered by evolution. In such a sequence, a number of proteins are induced to change their shape, which enables them to interact with each other so as to perform the task in hand. However, the term *signaling cascade* is now so entrenched in the literature that its continued use is assured, but we should think of the machinery as well as the information content of these signals.

The great complexity of the biochemical processes underlying synaptic plasticity is vividly illustrated in three recent reports. The first was by Hevroni et al. (1998), who stimulated glutamate receptors in the dentate gyrus with kainic acid and measured the resulting changes in mRNA levels. They found an increase in 362 mRNA levels and a decrease in 41. They identified 71 of these mRNAs as involved in a variety of signal transduction processes, tropic factors, vesicle proteins, structural and related proteins, retrograde messenger enzymes, protease inhibitors, phosphatases, kinases, receptors and channels, and neurotransmitters (NT) and their enzymes (see appendix A).

The second report was by Husi et al. (2000) (see also Sheng and Lee 2000). This group used a novel mass spectrometry technique to identify many of the proteins located in the postsynaptic density associated with glutamate *N*-methyl-D-aspartate receptors (NMDArs). They were able to identify seventy-one proteins (more than doubling the previously known number). This included five *N*-methyl-D-aspartate receptors, but no alpha-amino-3-hydroxy-5-methyl-4-isoxazoleproprionic acid (AMPA) receptors (AMPArs); nine scaffolding and adaptor proteins; ten phosphokinases; six phosphatases; two tyrosine kinases; nine mitogen-activated protein kinase (MAPK) pathway proteins; six small G-proteins and modulators; seven other signaling protein molecules; and sixteen cell-adhesion and cytoskeletal proteins.

The third report is by Craig and Boudin (2001), who claim that every synapse is biochemically unique and different from every other synapse—depending on presynaptic and postsynaptic cell types, environmental factors, developmental status, and history of activity.

Cerione (2000, p. 556) stresses "the staggering complexity that underlies cellular signaling events," which now requires "computational methods to integrate branching pathways that begin at the level of the cell surface receptors and continue at each subsequent downstream step. It is becoming impossible to conceptualize how all the cross-talk between different signaling pathways is achieved and worse, how the multiple and sometimes opposing outputs from a single stimulus are integrated to yield a final net cellular response. Cell and molecular biologists working on signaling problems will need help and lots of it from those trained in computational approaches" (Cerione 2000, p. 556).

Growing new spines and synapses is not simply a matter of reorganizing the cytoskeleton by promoting actin polymerization and supplying new membrane. All the rest of the complex cellular machinery of the synapse must be provided as well. This includes the cell recognition and adhesion molecules (CAMs) that ensure that the growing axon terminal contacts the correct spine and stays there. Also, CAMs, as we will see, need to be provided with their own signaling systems. Then a new postsynaptic density must be provided with its very complex system of scaffolding proteins that organizes the postsynaptic cascades of signaling molecules and maintains their contacts with membrane-bound receptors in one direction and the cytoskeleton in the other.

The new systems to be provided include the mechanism that mediates endocytosis of receptors and the external membrane as well as the system that sends signals to the nucleus and mediates transport of the resulting new mRNAs to their proper targets. Finally, there is a wide range of specific enzymes, channels, tropic factors, etc. that are necessary for the proper functioning of the new synapse. All these new components of the cellular machinery must be provided in a coordinated manner for the formation, maintenance, and final removal of the new synapse.

Some of these new proteins and membrane can be provided by local shunting from internal stores or from areas of synapse pruning. Others will be provided by new protein synthesis either at the level of the nucleus or locally from mRNAs in the dendrites.

It might be thought possible that the new external membrane required by the growing synapse might be provided by simply stretching the old membrane. However, biological membranes can only expand by <3% before rupturing (Nichol and Hutter 1996). Moreover, there is direct evidence that when a cell swells, the new membrane is provided from internal stores (Dai et al. 1998; Mills and Morris 1998).

1.2 Evidence for Synaptic Plasticity

The experiments that first suggested synaptic plasticity were carried out to demonstrate the effect of environmental conditions on dendritic growth (Quartz and Sejnowski 1997). To give a few examples out of a vast literature, mice reared in the dark and then placed in a lighted environment develop new apical dendrites in the visual cortex within 2 days (Valverde 1971). Animals reared in a complex environment develop more dendritic branching than do animals reared in a simple environment (Volkmar and Greenough 1972; Uylings et al. 1978; Fiala et al. 1978; Juraska 1980; Camel et al. 1986; Wallace et al. 1992). Rats reared in a complex environment show a 60% increase in multiheaded spines in the striatum compared with rats reared in a simple environment (Comery et al. 1996). Rats reared with much activity in a complex environment show an increase in dendrites and synapses in the cerebellum and cortex. This was not linked to mere increased activity, but to active maze learning, motor skill learning, and acrobatic skills (Kleim et al. 1996, 1997).

In songbirds, the area of brain that controls song increases significantly at the start of the breeding season. This involves an increase in the number, size, and spacing of neurons modulated by the increase in sex hormones (Tramontin and Brenowitz 2000).

Quartz and Sejnowski (1997) point out that synaptic plasticity is not simply a matter of an increase or a decrease in the number of synapses in the system. The precise location of these synapses on the dendritic tree, as well as the actual geometry of these branches, is also significant. In the nervous system of the blowfly, the shape of the dendritic tree of a visual neuron defines the region of the visual field within which that neuron is most sensitive to motion (Laughlan 1999). Quartz and Sejnowski (1997) also stress that local dendritic segments rather than entire neurons might constitute the brain's basic computational units. They add that it is significant that in these units dendritic membranes have nonlinear properties.

1.3 An Introduction to the Glutamate Synapse and Its Redox-Related Biochemical Features

A good place to start unraveling this puzzle is the glutamate synapse. There are three types of receptors at the glutamate synapse: AMPArs, NMDArs, and metabotropic glutamate receptors (mGlurs). Stimulation of the AMPAr opens an ion channel, which conducts mainly sodium ions (and some calcium) and mediates the principal depolarizing effects of glutamate on the postsynaptic membrane. However, it also activates the extracellular signal-regulating kinase (ERK-1/ERK-2) microtubule-associated protein (MAP) kinase in the hippocampus, leading to autophosphorylation of threonine-183 (Thr-183) and tyrosine-185 (Tyr-185) (Brown and Bahr 2000). Stimulation of the NMDAr opens a Ca^{2+} channel. However, it does this only when the membrane is partly depolarized, since the lumen of the channel is normally blocked by an Mg^{2+} ion. The prior depolarization expels this Mg^{2+} ion. Metabotropic glutamate receptors are G-protein-linked receptors that do not control ion channels but initiate various postsynaptic cascades.

Exocytosis of one presynaptic vesicle is sufficient to saturate all postsynaptic glutamate receptors (Conti and Weinberg 1999). The synaptic activity is terminated within 1 ms by rapid uptake of glutamate by the glutamate transporters (GluTs) located on neighboring astrocytes. In the astrocyte, the glutamate is metabolized to glutamine, which is exported to the glutamate terminal and there reconverted to glutamate. The calcium ions that enter the postsynaptic neuron following the opening of the NMDAr-regulated channel start a number of cascades important for synaptic plasticity, which I describe further later. Many of these cascades involve redox mechanisms.

Redox mechanisms play an important role in the glutamate synapse. Activation of glutamate synapses in the cortex, but not the cerebellum, leads to the release of ascorbate (vitamin C), which is the principal extracellular antioxidant in the brain (Rebec and Pierce 1994) (figure 1.1). When the glutamate transporter removes glutamate from the synapse, ascorbate is released into the synapse (Grünewald, 1993; Rebec and Pierce, 1994). The underlying mechanism may be competition between ascorbate and glutamate for a common binding site at the presynaptic site (Grünewald, 1993). Another suggested mechanism is that glutamate in the synapse activates ascorbate-permeable volume-sensitive organic anion channels in the astrocyte external membrane and so triggers ascorbate release into the synaptic cleft in this manner (Wilson et al. 2000a).

Glial cells actively release glutathione and the antioxidant enzyme, superoxide dismutase (SoD), into the synaptic cleft (Stone et al. 1999).

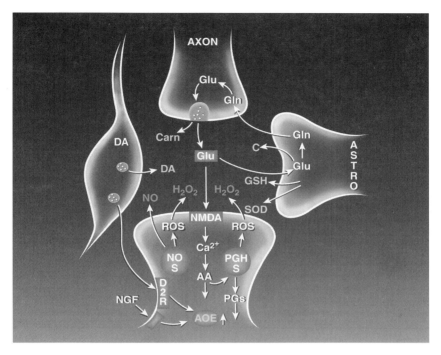

Figure 1.1
Redox aspects of the glutamate synapse. DA, dopamine; Carn, carnosine; Glu, glutamate; Gln, glutamine; GSH, glutathione peroxidase; SOD, superoxide dismutase; NMDA, *N*-methyl-D-aspartate; ROS, reactive oxygen species; PGH, prostaglandin H; PG, prostaglandin; D2R, dopamine receptor; AA, arachidonic acid; NGF, nerve growth factor; AOE, antioxidant enzymes; and ASTRO, astrocytes.

Glutathione is the principal intracellular antioxidant in the brain. The presynaptic glutamate vesicles also contain the antioxidant polypeptide, carnosine, which is released into the cleft together with the glutamate (Boldyrev et al. 1997).

The presence of these antioxidants in the synaptic cleft suggested (Smythies, 1997) that there should also be oxidants in the cleft, because a major function of antioxidants is to neutralize neurotoxic oxidants. An examination of the postsynaptic cascade following activation of the NMDAr revealed two such sources. The Ca^{2+} transmitted by the open NMDAr channel activates phospholipase A2, which converts membrane lipids into the second messenger, arachidonic acid (AA). The AA in turn acts as a substrate for prostaglandin H (PGH) synthase, which is the rate-limiting enzyme for prostaglandin synthesis.

PGH synthase releases large quantities of reactive oxygen species (ROS) such as superoxide[3] and hydrogen peroxide during its enzymatic activity.

The calcium inflow also activates nitric oxide synthase, another enzyme that produces ROS as a by-product of its activity. NO itself is also usually a pro-oxidant via its predominant form, the nitric oxide radical. However, under certain circumstances it can also act as an antioxidant (Chiueh and Rauhala 1999). Both hydrogen peroxide and nitric oxide are molecules that can diffuse freely through neurons, membranes as well as cytoplasm. They can therefore act as retrograde messengers to the synaptic cleft.

Many glutamate synapses have closely attached to one side a dopamine bouton-en-passage (Kötter 1994) so that the dopamine released can enter the glutamate synapse. All catecholamines, including dopamine, are potent antioxidants and free radical scavengers (Liu and Mori 1993). Much learning depends on positive reinforcement. It has been claimed that the widespread and diffuse volume release of dopamine in the brain carries the signal "positive reinforcement received" (Smiley et al. 1994; Dismukes 1977; Pickel et al. 1997; Rebec et al. 1997; Schultz 1998). More precisely, dopamine release is stated to increase in the prefrontal cortex and nucleus accumbens (but not in the striatum) (Rebec et al. 1997) if the degree of positive reinforcement received is greater than what the brain had computed it was likely to receive. Conversely, the amount of dopamine released is stated to diminish if the reinforcement received was less than expected (Schultz 1997).

Rebec et al. (1997) conclude that there is ample evidence for the role of dopamine neurons in reward signaling. However, a caveat is entered by Horvitz (2000), who says there is recent evidence to suggest that dopamine is released following stimuli other than rewarding ones—for example, stimuli mediating appetite, aversion, high-stimulus intensity, and novelty. He concludes that this means that dopamine release cannot simply encode reward. Spanagel and Weiss (1999) also argue that dopamine is not simply a signal for reward, but is involved in the formation of associations between salient contextual stimuli and internal rewarding or aversive events. In other words, they suggest that dopamine highlights "important" stimuli. In either case, a dopamine release signaling either a reward or important signals is compatible with my suggestion that the antioxidant effects of dopamine contribute to synaptic potentiation.[4]

Therefore I put forward the hypothesis (Smythies 1997) that one factor (out of many) that modulates the growth and pruning of synapses is the redox balance between oxidants and antioxidants at the glutamate synapse. Tilting this balance in the antioxidant direction would promote synaptic growth. Tilting it in the oxidant direction would lead to synaptic pruning. The amount of the antioxidant dopamine released could thus play a key role in this balance, and this would explain its role in learning. The question then arose of how the redox balance at the synapse could modulate synaptic plasticity.

The first answer I came up with was that the neurotoxic oxidants simply oxidized proteins and lipids in the synapse, leading to its destruction (Smythies 1997). This is the mechanism by which polymorphonuclear leukocytes destroy invading pathogens. The corpus luteum is removed at the end of pregnancy by the activity of ROS. This may include oxidative attack and/or apoptosis (Kato et al. 1997). However, as Ma et al. (1999) point out, it is easier and cheaper to switch off synapses than to remove them altogether. This led to the concept of "silent" synapses. These are synapses that possess only NMDArs but that can, following increased synaptic stimulation, recruit AMPArs to become fully operative synapses.

However, intermediate between physical damage of the synapses mediated by oxidative attack and switching them off by removal of AMAPrs, there is a third alternative. The recent advances in cell biology shortly to be described have revealed that the neuronal membrane is subject to a continual dynamic process of endocytosis into the postsynaptic neuron, processing by the endosome system, and recycling to the surface. One function of this process may be to redistribute membrane from areas of spine removal to areas of spine growth. There may also be changes in the overall dynamic balance between membrane endocytosis and recycling. There is evidence that ROS play a role in this process.

A fourth alternative is that ROS affect synaptic plasticity by an influence on a number of biochemical mechanisms involved in synaptic plasticity. I will return to this subject later. I now discuss the dynamic aspects of neurons in more detail.

1.4 Biochemical Factors in Synaptic Plasticity

An enormous effort has been expended in the past few years on discovering the basis in cell biology and biochemistry for synaptic plasticity. Glutamate is the neurotransmitter found at most excitatory synapses in the brain, and most glutamate receptors are located on dendritic spines. Therefore attention has been focused on the relationship between glutamate and spine numbers, the growth and shapes of spines, and the weighting of glutamate receptors.

In general, activity at glutamate synapses is necessary for synapses to grow or to remain in existence. Excess glutamate is, however, neurotoxic and can lead to spine destruction. Glutamate denervation can also lead to compensatory spine growth in the denervated region.

Rutledge et al. (1974) stimulated the cat suprasylvian gyrus for 2 seconds twenty times a day for several weeks. In some cases (A) the stimulus was paired with foot shock. In other cases (B) it was not. They then examined the spines and dendrites on the pyramidal cells in contralateral cortical layers II and III. In case (A) the dendrites had more terminal branches, were longer, and had more spines than controls. In (B) there was no such effect. Thus they concluded that spine and dendrite growth and elaboration were related to the learning rather than to mere stimulation.

Wilson et al. (2000b) applied glutamate to neurons in tissue culture and found a biphasic effect. In the short term (1 hour), this led to a calmodulin-associated increase in the growth rate of dendrites. However, during the next 24 hours a calpain-associated dendrite retraction occurred. The calcium-binding protein, calmodulin, binds directly to many cytoskeletal proteins and also activates various kinases and phosphatases, which would have an indirect effect on the cytoskeleton. Calpain is a calcium-activated protease that preferentially degrades cytoskeletal proteins.

Changes in intracellular calcium levels can regulate calmodulin in three ways.

1. At the cellular level by directing its subcellular distribution
2. At the molecular level by promoting different means of association with its target proteins
3. By directing a variety of conformational states of the calmodulin molecule that result in target-specific activation (Chin and Means 2000)

A combined lesion of the entorhinal cortex and fimbria or fornix deafferents the dentate gyrus (Schauwecker and McNeill 1996). This led to a decrease in dendritic length and branching within 4 days (and complete recovery by 45 days). There was also a 60% decrease in spines at 4 days (and an 87% recovery in 30 days). These authors also suggested that if spines are lost (e.g., during aging), the remainder could become functionally stronger to compensate. Spine numbers and dendritic branching decline with age. This underscores the importance of a life-long commitment to a cognitively invigorating program (Jacobs et al. 1997).

In a series of experiments carried out by McKinney et al. (1999), AMPAr antagonists led to a 50% loss of spine numbers. NMDAr antagonists did not affect spine numbers, but many of these changed into primitive filamentous filopodia. Spines are also not maintained in the absence of miniature synaptic potentials (MSPs). McKinney et al. (1999) suggested that the function of these MSPs is to prevent spine retraction during periods of relative inactivity at that spine. Frotscher et al. (2000) showed that in the case of spine growth in fascia dentata granule cells, innervation by fibers from the entorhinal cortex, but not neuronal activity, is essential for the normal development of granule cell dendrites; however, neuronal activity is required for the maturation of dendritic spines.

The effects of glutamate-related compounds have also been examined on cultured neurons. The effect of glutamate is dose dependent. A small dose causes spine elongation and a large dose results in spine shrinkage (Korkotian and Segal 1999) or loss (Halpain et al. 1998). Likewise, a small dose of NMDA causes neurites to grow, whereas a large dose kills the neuron (Dickie et al. 1996). Large doses of AMPA (but not mGlur agonists) are also neurotoxic (Halpain et al. 1998). NMDA and AMPA are more neurotoxic than glutamate because they do not activate protective group 1 mGlurs, which glutamate does (Cambonie et al. 2000).

Activation of NMDArs in situ by tetanic stimulation leads to the formation of new spines as early as 3 minutes (Engert and Bonhoeffer 1999). In CA1 neurons in the hippocampus, activation of NMDArs leads to growth of new filopodia (primitive spines) within minutes (Maletic-Savatic et al. 1999). Long-term stimulation of the ventroposterolateral (VPL) thalamic nucleus leads to a 25% increase in synapses in layers II/III of the motor cortex (Keller et al. 1992). Kossell et al. (1997) have studied in culture the effects of afferent stimulation on spines and dendrites. In preparations lacking afferent innervation, dendrites do not develop branches. In preparations having an afferent innervation, if this is not stimulated, the dendrites develop branches, but no spines. Only when the afferent inflow is activated do spines develop.

Fischer et al. (2000) report that activation of glutamate receptors reduces actin mobility in spines. Activation of NMDArs leads to the outgrowth of mobile spines from dendrites. Established spines need continual AMPAr activation to survive. AMPAr activation stabilizes already-formed spines independently of NMDAr activity (Fischer et al. 2000).

Scott and Luo (2001) have distinguished different factors that are concerned with different stages of dendritic growth. In the case of dendritic guidance, semaphorin-3A and the gene enabled are involved. In dendritic branching, the small guanosine 5′-triphosphatase (GTPase) Rac and the kakopo gene (which codes for a cytoskeletal linker protein that joins actin and microtubules) are important. Rac is also involved in spine formation. The growth of dendrites is limited by specific mechanisms that use the flamingo gene (which codes for a member of the cadherin family of CAMs), the small GTPase Rho A (via the rho-associated kinase, ROCK), and the Notch receptor.

NMDA and dopamine receptor activation leads to an increase in the production of the transcription factor Fos in the limbic system and basal ganglia by regionally differentiated but interdependent mechanisms (Radulovic et al. 2000). Hisanaga et al. (1992) report that NMDAr activation is required for c-Fos mRNA translation after stimulation of multiple intracellular signaling pathways, which further extends the influence of stimulation of NMDArs on cell mechanisms.

On the other hand, Rocha and Sur (1995) found that NMDA antagonists increase (sixfold) spine density and dendritic branch points on visual thalamic neurons in the lateral geniculate nucleus after a few hours in brain slices. These authors suggest that an active cellular mechanism involving phosphorylation and leading to the formation of dendritic spines is negatively regulated by afferent activity. NMDAr activation also leads to a diminution of the normal proliferation of dentate gyrus neurons (Gould and Cameron 1997). In all these experiments, the quantity of glutamate effective at the synapse, as well as possible microanatomical considerations, is probably critical in determining whether the chemical stimulation results in synapse deletion or growth.

1.5 The Biochemical Basis of the Hebbian Synapse

The Hebbian synapse is one whose efficiency is increased if its activation occurs within a critical time frame of 20 ms before depolarization of the postsynaptic membrane. The NMDAr is thought to play a key role in this, as detailed earlier. Most studies in this field have been carried out on neuronal targets subjected to high-frequency tetanization, which induces long-term potentiation (LTP). Low-frequency tetanization induces long-term depression (LTD). During LTP, the synapse responds more efficiently to a fixed stimulus. However, high-frequency tetanization is a rare feature in vivo and possibly a better model for the circumstances that evoke a Hebbian synapse in vivo may be backpropagation (Paulsen and Sejnowski 2000; Williams and Stuart 2000). The action potential is generated by the axon hillock of a neuron in response to dendritic depolarization that is mediated mainly by activation of AMPArs. The action potential is then sent down its axon, but by backpropagation, it also invades the soma and dendrites carried by voltage-activated sodium channels. This depolarization will unblock NMDAr channels within the necessary time window and will also lead to the activation of voltage-activated calcium channels. The resulting inflow of calcium into the postsynaptic cell lays the groundwork for the Hebbian effect (Linden 1999). Paulsen and Sejnowski (2000) stressed the role of burst firing of the postsynaptic neuron in promoting synapse growth. They distinguish three levels of signaling in memory encoding:

(a) Silence
(b) Single spike firing that transfers information
(c) Burst firing that signifies changes in synaptic weights

Burst firing in hippocampal neurons occurs in two circumstances: (1) during the active exploration of a novel environment, those neurons that code for the current locus show bursts of activity and (2) in slow wave sleep, bursts occur during the replay of spike frequencies related to events of the previous day. In developing neurons, (b) is sufficient for laying down neural architecture, but in adult neurons, (c) is required.

Paulsen and Sejnowski (2000) also codify the Hebbian rule in this context: Those synapses are potentiated that are active immediately preceding the postsynaptic spike. Those synapses that are activated directly after the postsynaptic spike are downregulated.

Since glutamate is present in the synapse for only 1 ms after it is released from the presynaptic terminal (Diamond and Jahr 1997), this suggests that when it binds to the NMDAr protein, it must in some way first prime that receptor without actually opening the channel. This is because the channel can only be opened when the membrane is depolarized to remove the blocking magnesium ion in the channel. Depending on where that particular spine is located on the dendritic tree, there will always be a delay of several milliseconds between the time when the NMDAr binds a molecule of glutamate and the time when the backpropagated action potential (BP-AP) gets back to that particular spine. For this mechanism to work, therefore, it would appear necessary that the NMDAr should not be rapidly endocytosed as soon as it binds the molecule of glutamate. This may be why the NMDAr is one of the few receptors that are not endocytosed upon activation. The reason for this resistance to endocytosis may be that the NMDAr is tightly bound to the postsynaptic density (PSD), whereas the AMPAr is only loosely attached (Gomperts et al. 2000).

The result of activating NMDArs depends in part on their location. In the optic tectum, a blockade of NMDArs leads to a change in the arbor dynamics of axons (an increase in branch additions and a decrease in branch lifetimes). However, this same blockade of NMDArs results in a decreased rate of dendritic branch tip additions and subtractions (Rajan et al. 1999). Rajan et al. suggest that the function of this is to increase the probability of coactive afferents converging on a common target during NMDAr inactivity. This means that during periods of relative NMDAr inactivity, the postsynaptic dendrites lower their rate of branch additions and so reduce the target for afferent input. The axons react by looking for new partners, increasing the number of short branches and reducing their lifetimes.

During periods of NMDAr activity, the aberrant axonal branches are retracted and the dendrites form more branches. The new synapses at first are "silent" in that they have only NMDArs; they gradually add AMPArs as they mature if they are activated. New axons that fail to activate NMDArs and thus fail to receive a retroactive stabilization signal are retracted. "Silent" synapses have been detected in neonatal rat visual cortex. In these experiments, following pairing of presynaptic stimulation and postsynaptic depolarization, there is a long-lasting induction of AMPArs and their trafficking to the membrane to form a functional synapse (Rumpel et al. 1998).

Growing retinal axons add new branches in vast excess, most of which are removed within 10 minutes (Rajan et al. 1999). It has been suggested that new synapses are "silent" in order to avoid disrupting already functioning networks to an undue degree (Feldman and Knudsen 1998). Quartz and Sejnowski (1997) have stipulated that the grafting of new synapses into old networks requires that two conditions be satisfied:

- The locality condition, in which the addition of a structure must be at the appropriate local scale and not result in wholesale changes in representation with each new elemental change
- The stability condition, in which under normal conditions local changes must not undo previous learning

Although LTP and LTD depend on using somewhat unphysiological methods of stimulating neurons, nevertheless some of the findings may turn out to be relevant to synaptic plasticity in the real brain. LTP requires both NMDAr and mGlur activation (Riedel and Reyman 1996). Hedberg and Stanton (1996) add a detail that this is the case for LTP production in monosynaptic systems, but LTP production in polysynaptic systems needs only NMDAr activation. Using direct vision of CA1 neuronal spines, it has been reported that LTP stimulation produces new spines in 30 minutes (Engert and Bonhoeffer 1999).

In contrast, Harris (1999) and Rusakov and Kullman (1998) claim that LTP induction in the dentate gyrus is associated with a 20% reduction in spine numbers and shorter, thicker, spines. Moreover, large changes in LTP can occur without any change in spine numbers or morphology (Collin et al. 1997). In the initial stage of new spine formation, LTP induces "perforated synapses." These are discontinuities in the PSD, which have higher than normal levels of AMPArs and smooth endoplasmic reticulum (ER) and are more likely to contain spine apparatus (Lüscher et al. 2000).

A number of factors have been correlated with LTP production, e.g., hydrogen peroxide (Katsuki et al. 1998), NO (Lu et al. 1999), arachidonic acid (Horimoto et al. 1997; Nishizaki et al. 1999), interleukins (Li et al. 1997; Balschun et al. 1998), muscarinic receptors (Centonze et al. 1999), and mGlurs (Grover and Yan 1999). These findings are best considered under their own headings in view of the possibly dubious relevance of LTP and LTD in real cell biology. Moreover, Holscher (1997) concludes that LTP and LDP are not reliable models for learning. Other critical evaluations of the relevance of LTP to memory formation have been made by Shors and Matzel (1997). However, recently evidence has been obtained (Rioult-Pedotti et al. 2000; Martin and Morris 2001) that some of the biochemical mechanisms involved in LTP are involved in real learning.

1.6 More Redox Reactions at the Synapse

The redox state of a biochemical system has many effects in neurons besides oxidative attack on proteins and lipids by oxidants (ROS) (such as the hydroxyl radical, hydrogen peroxide, and superoxide), resulting in neurodestruction and defense against this attack by antioxidants (such as ascorbate, glutathione, and vitamin E). The sources of ROS in cellular metabolism include:

- Several mitochondrial enzymes in the electron chain (mitochondria convert 5% of the oxygen they consume into superoxide)
- Xanthine oxidase
- Monoamine oxidase
- Cyclo-oxygenase
- Nitric oxide synthase
- Lipoxygenases

ROS themselves have a wide variety of effects on biological systems:

1. ROS have five roles in signal transduction. They affect cytokines, growth factors, and the secretion and action of hormones; ion transport; transcription; neuromodulation; and apoptosis.

2. They modulate the performance of various enzymes, e.g., tyrosine phosphatases (down), serine/threonine (Ser/Thr) phosphatases (down), and Ser/Thr kinases (up) (Kamata and Hirata 1999).

3. Oxidants cause a rise in intracellular calcium levels by modulating a number of regulators, such as Na^+/K^+ and Na^+/Ca^{2+} exchangers, Na^+/K^+-adenosine triphosphatase (ATPase), Ca^{2+}-ATPase, and various calcium channels. This rise in intracellular calcium leads to the induction of various protein kinases and increased levels of phosphorylation. Oxidants thus mimic the effects of stimulation of the NMDAr by raising intraneuronal calcium levels (Chakraborti et al. 1998).

An important paper by Yermolaieva et al. (2000) provides evidence of the direct effect of ROS on synaptic plasticity. In rat cortical brain slices, paired application of agonist stimulation and ROS resulted in a long-lasting increase in calcium signaling (which was reversed by hypoxia). This increase critically depended on NO production. These authors suggest that ROS play a critical role in calcium homeostasis via oxidation of amino acid residues in proteins, e.g., on intracellular Ca^{2+} release channel ryanodine receptors and methionine residues on calmodulin. The accessory action of NO might be due to upregulating a cyclic guanosine 5'-monophosphate (cGMP) signaling pathway or to its direct free radical action (Yermolaieva et al. 2000).

4. Oxidative stress activates transcription factor nuclear factor-κB (NF-κB), which leads to the increased production of Mg^{2+} superoxide dismutase and decreased synthesis of peroxynitrite (Mattson et al. 1997). This represents a neuroprotective feedback system. In glia, NO leads to decreased NF-κB function. The mechanism may act by inducing and stabilizing the NF-κB inhibitor IκB-α or by inhibiting the binding of NF-κB to DNA by S-nitrosylation of cysteine-62 (Cys-62) of the p50 subunit (Colasanti and Persichini 2000). Hutter and Greene (2000) give a comprehensive review of the multiple factors by which redox states influence gene expression, in particular via NF-κB. Janssen-Heininger et al. (2000) give an extensive review of the role of redox factors in modulating NF-κB function. They point out that important factors involved are the site of generation of the oxidants, the type of oxidant, and the time frame.

5. Chaperone proteins are concerned with ensuring the correct folding of proteins. This reaction is subject to redox modulation. Oxidative stress oxidizes sulfhydryl groups to disulfide bonds, which activates the chaperone function (Jakob et al. 1999).

6. The reduction in protein synthesis induced by activation of NMDArs is mediated by ROS (Monje et al. 2000).

7. Ravati et al. (2000) have found that a short and moderate exposure to oxidative stress produced by several agents (e.g., glutamate, staurosporine, ischemia, anoxia) that use ROS as a final common path protects against later severe damage by ROS. They called this effect "preconditioning." In the brain it is mediated in part by upregulation of heatshock proteins, opening of ATP-sensitive K^+ channels, and

upregulation of the synthesis of antioxidant enzymes such as Mn^{2+} SOD. These authors conclude that there is "increasing evidence that ROS are important intracellular signaling molecules modulating the phosphorylation status of several proteins that are important for cellular integrity" (Ravati et al. 2000, p. 31).

8. These reports that ROS can upregulate kinases and downregulate phosphatases suggests that the basic redox theory needs modification. Upregulating kinases and downregulating phosphatases is usually associated with synaptic promotion, not removal (Coussens and Teyler 1996; Tokuda and Hatase 1998).

On the other hand, the redox state also affects activation of transcription factors, protein conformation, and metabolism of calcium. Any of these can activate an isoconverting enzyme–like neurodestructive protease cascade (Clément et al. 1998). Activation of such a cascade might be expected to promote spine pruning. As Sen (1998, p. 1747) has said, "Redox-based regulation of signal transduction and gene expression is emerging as a fundamental regulatory system." This is echoed by Kamata et al. (2000), who state that the cellular redox state has been shown to play an essential role in cellular signaling systems. Thus in certain microanatomical situations ROS could act to promote synaptic plasticity, whereas in other microanatomical situations they could have the opposite effect. One role of ambient antioxidants might be to guard against the neurotoxicity induced by excess ROS production. But they might have many relevant effects in their own right. The action of ROS and antioxidants on synaptic plasticity may depend on the summation and integration of effects on a large number of different relevant biochemical systems.

9. ROS are also involved in mediating cellular responses to changes in the ambient oxygen concentration (Kietzmann and Fandrey 2000). The oxygen sensor is a heme oxidase [e.g., reduced nicotinamide adenine dinucleotide phosphate (NADPH) oxidase]. Low ambient oxygen levels depress hydrogen peroxide production by this enzyme and high ambient oxygen levels increase it. The hydrogen peroxide diffuses to the neighborhood of the gene, where a Fenton reaction takes place by interaction with iron-producing hydroxyl radicals. These oxidize sulfhydryl groups in certain candidate transcription factors. The latter now bind to gene elements that produce proteins that are desirable in a high oxygen environment (Kietzmann et al. 2000).

10. In some cases superoxide signaling may be mediated by its interaction with NO to produce peroxynitrite. Ullrich and Bachschmid (2000) argue that peroxynitrite may not merely be the villain that it is often portrayed to be, with a role restricted to pathological reactions, but at low levels may have some normal physiological roles.

1.6.1 Hydrogen Peroxide

Hydrogen peroxide is a particularly important ROS because of its ability to diffuse freely through cellular tissue. It is produced by the action of numerous enzymes and is degraded by glutathione peroxidase in the cytosol and mitochondria, and by catalase in perisomes. Furthermore, when it makes contact with free iron, it is degraded to form the highly toxic hydroxyl radical by the Fenton reaction. Hydrogen peroxide has a number of effects.

Effects on Neurotransmission   Hydrogen peroxide reduces GABAergic (gamma-aminobutyric acid) inhibition in the cortex and thalamus, probably by impairing chloride gradients (Sah and Schwartz-Bloom 1999). In the thalamus it also potentiates glutamate excitation. It raises the excitability of cortical neurons only if the thalamic input to the cortex is preserved. This is mediated by continuous neuronal firing and long depolarization shifts in response to a stimulus. In effect, hydrogen peroxide converts a low-pass filter to a flat broad-pass filter (Frantseva et al. 1998).

It blocks catecholamine uptake into synaptic vesicles (dopamine $\gg$ norepinephrine) (Langeveld et al. 1995) and blocks glutamate uptake into vesicles by inhibiting the proton ATPase in the membrane (Wang and Floor 1998).

It induces an adenosine-mediated decrease in synaptic transmission in hippocampal slices (Masino et al. 1999).

It inhibits synaptic dopamine release, possibly via reduction of ATP production, oxidation of SNARE (soluble N-ethylamide-sensitive fusion protein attachment protein receptor) protein, and/or increased phosphorylation of calcium-binding proteins (Chen et al. 2001).

Effects Mediated by Changes in Calcium Levels   Hydrogen peroxide's ability to raise intracellular calcium levels and to lower glutathione levels renders it a potent and effective neurotoxin (Hoyt et al. 1997). The raised Ca^{2+} levels lead to increased binding of annexin V to cell membranes, which can lead to apoptosis (Oyama et al. 1999). The raised Ca^{2+} levels also activate calpain, leading to degradation of cytoskeletal proteins (Ishihara et al. 2000). Apoptosis is also induced by hydrogen peroxide by a secondary stimulation of NMDA receptors that occurs after hydrogen peroxide washout. The major part of hydrogen peroxide–induced neurotoxicity is due to hydroxyl radical formation. This may be related to the delayed accumulation of extracellular glutamate and NMDA receptor activation and to poly(adenosine diphosphate ribose) (ADP ribose) polymerase activation and the related decrease in nicotinamide adenine dinucleotide (NAD) content. The combination of these two mechanisms would lead to both an increase in adenosine triphosphate (ATP) consumption and a decrease in ATP synthesis, leading to cell death (Mailly et al. 1999).

Hydrogen peroxide mediates the effect of epidermal growth factor (EGF) in raising intracellular calcium levels (via Rac and RhoA) (Lee et al. 2000). Kamata et al. (2000) have examined this system in detail and find that hydrogen peroxide regulates the EGF receptor by multiple steps, including dephosphorylation by protein tyrosine phosphatase, ligand binding, and a Cys adenylate cyclase (Ac)-sensitive cellular process. Hydrogen peroxide also modulates the signaling cascade initiated by EGF by action on the phosphorylation of extracellular signal-regulated kinases 1 and 2 and of cyclic adenosine monophosphate (cAMP) element-response binding protein (CREB) (Zhang and Jope 1999). These authors point out that this may be of relevance to Alzheimer's disease because CREB is involved in the formation of long-term memory.

Hydrogen peroxide inhibits Ca^{2+}-dependent glutamate release without affecting cytosolic calcium levels (Zoccarato et al. 1999).

In cortical brain slices from young rats, agonist-induced depolarization paired with oxidation induces long-lasting potentiation of subsequent Ca^{2+} signaling that is reversed by hypoxia. This potentiation provides direct evidence of the role of redox factors in synaptic potentiation. It critically depends on NO production and is mediated by ROS utilization. This effect decreases with aging (Yermolaieva et al. 2000).

Effects on Second Messengers   Hydrogen peroxide stimulates spingomyelin hydrolysis to yield ceramide. Ceramide is normally a second messenger for cytokine stimulation and inhibits ionotropic glutamate transmission by activating postsynaptic protein phosphatases-1 and 2A (PP-1 and PP-2A). It can also activate protein kinase ζ and MAPK (Yang 2000). Ceramide can act as an apoptotic agent by these pathways (Goldkorn et al. 1998). Ceramide and sphingosine also increase the cell's vulnerability to neurotoxicity induced by hydrogen peroxide (Denisova et al. 1999). A comprehensive review of ceramide function in cells has been given by Kolesnick et al. (2000).

Hydrogen peroxide increases phosphoinositide turnover, which leads to an important signaling cascade (Suzuki and Ono 1999; Suzuki et al. 1997). It also increases the accumulation of inositol phosphate (IP) (from phospholipase C, PLC) in both astrocytes and brain slices (Servitja et al. 2000).

Hydrogen peroxide affects phospholipid signaling by another mechanism—by increasing the formation of phosphatidic acid and the accumulation of phosphatidyl butanol (a product of phospholipase D, PLD) (Servitja et al. 2000).

Hydrogen peroxide alone does not affect arachidonic acid release, but it strongly potentiates the release of AA induced by NMDAr stimulation (Samanta et al. 1998).

Hydrogen peroxide activates c-Jun NH_2-terminal kinase (JNK). In this, Src kinase and its substrate Cas play an essential part (Yoshizumi et al. 2000).

Hydrogen peroxide potentiates gene expression by its effect on mRNAs and by modulating various transcription factors such as clathrin adaptor protein-1 (AP-1), NF-κB, and the actin gene enhancer CArG (Sakamoto et al. 1999).

It stimulates forskolin-generated cAMP production (Raimondi et al. 2000). The source of hydrogen peroxide in these studies was tyramine oxidation by monoamine oxidase in rat white adipocytes.

ROS activate the protein kinases ERK and MAPK (by upregulating their phosphorylation) and also activate ras and recruitment of Raf kinase to the plasma membrane, where it becomes activated (Mukhin et al. 2000).

Effect on Endocytosis   Hydrogen peroxide inhibits the endocytosis of EGF receptors by action in an early step in endocytosis, possibly ubiquitation (de Wit et al. 2000). These authors go on to suggest that receptor-mediated endocytosis might be inhibited in a general way by oxygen free radicals. Inhibition of endocytosis might be expected to impair new synapse formation and thus would be an example of how redox mechanisms affect synaptic plasticity.[5] The redox theory of synaptic plasticity (Smythies 1997) suggests that the redox balance at a synapse between neurotoxic oxidants (ROS and reactive nitrogen species, RNS) and neuroprotective antioxidants might modulate synaptic plasticity. In the original theory it was suggested that synapses might be pruned by oxidative attack on the proteins and lipids in the membrane. However, it is cheaper and more efficient if ROS were to prune synapses by inhibition of endocytosis (thus cutting off the supply of membrane for new synapses).

Miscellaneous Systems   In astrocytes, hydrogen peroxide reduces high-energy phosphate (ATP, GTP) levels and deregulates control of osmosis. It also promotes the pentose shunt pathway of glucose metabolism (Brand et al. 1999).

Atkins and Sweatt (1999) showed that diffusible ROS and RNS generated during increased activity of CA1 neurons in the hippocampus diffuse to neighboring oligodendrocytes and there induce post-translational modification of myelin basic protein.

In neutrophils, substance P activates calmodulin-dependent NADPH oxidase, which leads to the generation of superoxide anion and hydrogen peroxide. It is not known if a similar mechanism may operate in neurons (Sterner-Kock et al. 1999).

The pleiotropic cytokine transforming growth factor-β1 (TGF-β1) plays a key role in wound healing and organ fibrosis by upregulating the α-1 procollagen gene. Some of its actions are mediated by formation of hydrogen peroxide. Therefore hydrogen peroxide may be one of the mediators of the healing response (Domínguez-Rosales et al. 2000).

Evidence for the presence of the Fenton reaction (which converts hydrogen peroxide in the presence of iron into the toxic hydroxyl radical) in vivo is suggested by the report that H_2O_2-sensitive LY-S cells have a free iron level three times higher than H_2O_2-insensitive LY-S cells (Lipinski et al. 2000).

The stress-inducible 23-kDa protein, peroxiredoxin, initially found in macrophages, also occurs in oligodendroglia and Schwann cells, and to a lesser extent in axons and neuropils. Since this protein reduces hydrogen peroxide, these data suggest that it may play an important role in protecting against oxidative stress in the brain (Mizusawa et al. 2000).

Hydrogen peroxide inhibits the proteolytic action of ATP-activated proteosome 26S; this effect may modulate cell-cycle control and other important physiological functions (Reinheckel et al. 2000).

It reduces GABAergic inhibition in the cortex and thalamus, probably by impairing chloride gradients (Sah and Schwartz-Bloom 1999).

It potentiates gene expression by its effect on mRNAs and by modulating various transcription factors such as AP-1, NF-κB, and CAuG (Sakamoto et al. 1999).

Hydrogen peroxide increases phosphoinositide turnover, which leads to an important signaling cascade (Suzuki and Ono 1999; Suzuki et al. 1997).

Hydrogen peroxide activates phospholipase D2 in a reaction dependent on Ca^{2+} and phosphokinase C (Oh et al. 2000).

All or any of these effects of hydrogen peroxide could be relevant to its effects on synaptic plasticity.

1.6.2 Redox-Sensitive Sites on Proteins

Some proteins concerned in neurotransmission have redox sites that modulate their activity. These sites consist of two cysteine moieties, oxidation of which forms a disulfide bond and downregulates the activity of that protein. The NMDAr has a redox site that acts as a negative feedback control of the receptor channel. Enzymes, such as NO synthase and PGH synthase (cyclo-oxygenase), in the calcium-mediated post-NMDAr cascade, are potent sources of ROS production. These ROS activate the redox site on the NMDAr molecule and so downregulate it and slow down the production of ROS by the cascade (Aizenman et al. 1989). This mechanism protects against glutamate neurotoxicity, which is mediated in part by excess intracellular calcium levels and the resulting excess ROS production.

The dopamine D1 receptor (D1R) has a similar redox site, which is "exquisitely sensitive" to modulatory redox changes (Sidhu 1999). The glutamate transporter molecule has a redox site, but the AMPArs and mGlurs do not (Trotti et al. 1996). ROS inactivate the glutamate transporter by oxidizing its redox site (Agostinho et al. 1997).

1.6.3 Carnosine

This small dipeptide (alanine-histidyl) is copackaged with glutamate in synaptic vesicles and is released with it into the synaptic cleft (Boldyrev et al. 1997). It protects against excitatory cell death produced by excess glutamate (Boldyrev et al. 1999). It has several properties that may contribute to this effect:

- It is an antioxidant and chelator of free radicals (Boldyrev et al. 1997).
- It chelates divalent metal ions (Sassoè-Pognetto et al. 1993), and blocks the Cu^{2+} and Zn^{2+}-induced inhibition of NMDA- and GABA-mediated transmission in the olfactory bulb (Trombley et al. 1998).
- It inhibits aldehyde-induced cross-linkage of proteins (Hipkiss 1998; Hipkiss et al. 1997).
- It may also act as an altruistic suicide molecule. Carnosine is easily oxidized by ROS and this may spare other more important molecules from suffering the same fate (Boldyrev et al. 1999).

1.6.4 Nerve Growth Factor: Redox Factors

Nerve growth factor (NGF) also operates on the redox system of the neuron. It binds to its receptor TrkA in the external membrane and is rapidly endocytosed. In the endosome, the NGF molecule is released and is trafficked to the nucleus, where it modulates transcription. NGF rapidly diminishes the increase in NO synthase activity induced by activation of NMDArs and AMPArs. It does this by modulating the expression of nitric oxide synthase (NOS) at the nuclear level (Lam et al. 1998). It also decreases the production of ROS involved by activation of the MAP kinase pathway (Dugan et al. 1997) and elevates the activity of the antioxidant enzymes, glutathione peroxidase (GSH-px) and catalase (CAT), by making their mRNAs more stable (Sampath and Perez-Polo 1997). Another neurotropin—activity-dependent neurotropic factor—protects against oxidative stress by raising the rate of glucose and glutamate transport and inhibiting the impairment of these by ferrous iron and beta-amyloid (Guo and Mattson 2000).

Brandner et al. (2000) injected NGF into the brains of neonatal rats at days 12 and 13 postpartum. This improved performance compared with controls during learning tasks undertaken at 22 days (Morris navigation task) and at 6 months (radial arm maze test).

1.6.5 Nitric Oxide: Redox Factors

In the cerebral cortex there are two types of NO synthase-positive neurons that produce NO:

- Large GABAergic neurons deep in layer VI with a few in layers II–IV
- Smaller, more numerous GABAergic neurons in all cortical layers, but mainly in layers II and IV (Yan and Garey 1997)

NO synthase is present in large amounts in spines and proximal dendrites. The soma and proximal dendrites have very low levels. Most NOS^+ axons are GABAergic, but some are glutamatergic (Aoki et al. 1997). The detailed distribution of NOS^+ neurons in brain has been given by Egberongbe et al. (1994). NO is released by axons and soma, not just by axon terminals (Wiklund et al. 1997).

Nitric oxide can be either a pro-oxidant or an antioxidant, depending on circumstances (Rosenberg et al. 1999; Lancelot et al. 1995; Koppenol 1998; Rauhala et al. 1996). NO may occur as the nitric acid radical or the nitrosium ion. The former is predominant and is strongly oxidant; the latter is weakly antioxidant. NO modulates the induction of mRNAs for CAT (down), GSHpx (down) and Cu^{2+}/Zn^{2+} SOD (up). Since SOD produces hydrogen peroxide and CAT metabolizes it, upregulation of the former associated with downregulation of the latter will result in excess production of hydrogen peroxide.

NO combines with superoxide to form the highly neurotoxic molecule, peroxynitrite ($ONOO^-$). This is normally generated by microglia stimulated by cytokines or beta-amyloid. It can also be generated by neurons in response to excess glutamate transmission, following depletion of L-arginine or tetrahydrobiopterin, and as a result of mitochondrial dysfunction (Torreilles et al. 1999). There is evidence that $ONOO^-$ may be a signaling molecule in its own right via the activation of PLA2 and AA (Guidarelli et al. 2000). $ONOO^-$ modulates calcium influx into neurons in a complex manner. It increases calcium influx by opening P/Q-type and L-type voltage-dependent calcium channels, but decreases calcium influx at N-type voltage-dependent calcium channels (Ohkuma et al. 2001).

Nitric oxide influences a number of systems important at synapses:

It has been suggested that nitric oxide S-nitrosylates sulfhydryl H groups at the redox site of the NMDAr and thus downregulates it (Bains and Ferguson 1997; Lei et al. 1992). On the other hand, claims have been made that NO diminishes NMDA-induced currents by interacting with cations rather than with the redox state of the receptor (Dawson and Dawson 1996).

It increases GABAergic inhibition by increasing GABA release and by potentiating the effects of GABA on its receptor.

It has also been suggested that NO acts mainly postsynaptically and is not just a retrograde messenger to the presynaptic terminal—at least in hippocampal CA1 cells (Ko and Kelly 1999). However, evidence that NO also acts as an activity-dependent retroactive messenger that modulates axon arbor formation has been provided by Cogen and Cohen-Cory (2000).

NO directly activates the cyclo-oxygenase pathway in neurons (Salvemini et al. 1993) and in astrocytes (Molina-Holgado et al. 1995).

Through its activation of guanyl cyclase and by raising cGMP levels, it affects a number of events downstream in this cascade, for example, raising intracellular calcium levels, activating protein kinases and phosphodiesterases, modulating gene transcription and translation, affecting the release and reuptake of neurotransmitters, and directly opening ion channels (Szabó 1996; Bonfoco et al. 1996). Nitric oxide inhibits both xanthine oxidase (Fukahori et al. 1994) and cytochrome oxidase (Brown 1997).

As we have seen, local events during early LTP tag active synapses so that these synapses can use newly synthesized proteins during late LTP. NO plays a role in this. The gas diffuses to presynaptic terminals, where it plays a role in early LTP, and to postsynaptic terminals, where it plays a role in late LTP. The mechanism acts by activation of guanyl cyclase. The cGMP produced activates a protein kinase that acts in parallel with phosphokinase A to phosphorylate the transcription factor, CREB (Lu et al. 1999).

1.6.6 Glutamate Neurotoxicity: Redox Factors

Glutamate neurotoxicity is mediated by excess ROS and RNS produced in the post-NMDAr cascade, in part by increased PGH synthase activity (producing ROS) and in part by the increased levels of NO synthase activity (producing RNS and ROS) (Dawson et al. 1991). This leads to depletion of glutathione and mitochondrial dysfunction (Almeida et al. 1998). Mitochondrial dysfunction leads to lowering of ATP levels and depolarization of the external membrane. Normal membrane polarization depends on the energy provided by ATP. This depolarization leads to activation of NMDArs and excess calcium inflow and neurotoxicity (Schulz et al. 1997).

Psychological stress raises NO production and leads to lipid peroxidation (Matsumoto et al. 1999). At an early stage of glutamate neurotoxicity, Atlante et al. (2001) report that cytochrome oxidase (an ROS scavenger) is released from mitochondria as a defense mechanism. Later stages of glutamate neurotoxicity are associated with damage to mitochondria (Atlante et al. 2001).

1.6.7 Glutathione

The principal intracellular antioxidant in brain is the tripeptide, glutathione. It has, however, properties other than its antioxidant function that may be relevant to synaptic plasticity. Its molecule contains both glutamate and cysteine moieties. Thus glutathione can act at the glutamate receptor as an agonist (at low doses) and as an antagonist (at high doses). It can also act as a reductant at the redox site of the NMDAr molecule via its cysteine moiety (Janáky et al. 1999; Shaw and Salt 1997; Varga et al. 1997). Both these activities may be relevant to its neuroprotective effect. Glutathione also modulates the action of the transcription, factor NF-κB within a narrow critical dose range. Excess glutathione blocks signal transduction, whereas too little inhibits the capacity of NF-κB to bind to DNA (Lander 1977).

1.6.8 Ascorbate

Ascorbate is normally an antioxidant and free radical scavenger. It too has other effects that may be significant at the synapse. Ascorbate has little effect by itself on the activity of striatal neurons, but it modulates their response to glutamate stimulation. This effect is dose dependent; a low dose activates, a high dose inhibits. This effect may be mediated via the glutamate transporter system (Kiyatin and Rebec 1998). Ascorbate also inhibits dopamine binding and acts behaviorally as a neuroleptic (Rebec and Pierce 1994). Ascorbate inhibits Na^+/K^+-ATPase and dopamine-sensitive adenylate cyclase. It acts as a cofactor for dopamine-β-hydroxylase and stimulates the release of acetylcholine and norepinephrine (NE) from synaptic vesicles (Milby et al. 1981).

Karanth et al. (2000) have presented evidence that they believe supports a role for ascorbate as a neurotransmitter (or neuromodulator). The release of luteinizing hormone-releasing hormone (LH-RH) is controlled by the release of NO from NOergic neurons placed in juxtaposition to the LH-RH terminal in the hypothalamus. NO activation of glutamyl cyclase starts a cascade that runs increased cGMP → activation of cyclo-oxygenase → increased PGE_2 levels → increased leukotriene levels. All these act in concert to lever the extrusion of LH-RH secretory granules into the hypophyseal portal vessels for delivery to the pituitary. The NOS neurons are activated by glutamatergic and nor-epinephrinergic inputs. Ascorbate inhibits this cascade by a direct chemical reaction (scavenging) of NO. Karanth et al. (2000) suggest that ascorbate, which is present in high concentration in glutamate synaptic vesicles, acts as a feedforward inhibitory control of LH-RH release in this manner.

The antioxidant properties of ascorbate play a role in hibernation. In this state, cerebral blood flow falls by 90% and there is rapid reperfusion on arousal. Anoxia followed by reperfusion is normally a perilous condition since the reperfusion leads to a massive release of ROS. This mechanism is responsible for much of the brain damage caused by strokes. During hibernation, plasma levels of ascorbate rise threefold and cerebrospinal fluid (CSF) levels rise twofold. These rapidly return to normal following arousal.

The ascorbate-dehydroascorbate redox cycle also plays an important role in electron transfer reactions from protein thiols in the endoplasmic reticulum. These reactions are required for the proper folding of proteins by formation of disulfide bonds (Csala et al. 1999). In the endoplasmic reticulum, overload with protein or the presence of unfolded protein leads to calcium release. This in turn induces ROS release and NF-κB activation. This results in the transcription of inflammatory and immune genes (Pahl 1999). The chaperone protein, glucose-regulated protein-78 (GRP-78), provides a protective negative feedback loop in the endoplasmic reticulum. When activated by ROS, GRP-78 inhibits ROS production and promotes stabilization of mitochondrial function (Yu et al. 1999).

In some circumstances, however, e.g., in the presence of free iron, ascorbate can act as a pro-oxidant. Furthermore, its oxidized form, dehydroascorbate, is rapidly taken up by the glutamate transporter into cells where it will act as a pro-oxidant (Song et al. 1999). Recently it has been found (Simpson and Ortwerth 2000) that a major metabolite of ascorbate is L-erythrulose, which is an extremely reactive ketose capable of glycating and cross-linking proteins. Thus, under certain circumstances ascorbate could do more harm than good.

1.6.9 Thioredoxin

Rybnikova et al. (2000) have described a novel antioxidant—thioredoxin-2—that is widely distributed in the brain. It has the redox-sensitive site -tryptophan-cysteine-glycine-proline-cysteine- and acts as a scavenger of ROS and a redox regulator. These authors suggest that thioredoxin-2 is an important antioxidant protectant in the brain.

Chapter 2

Endocytosis and Exocytosis

2.1 The Role of Endocytosis

Over the past few years discoveries made in cell biology have gradually seeped over into neuroscience. A major example of this is the role of endocytosis in neuronal function. The mechanism responsible for this function is as follows: In the living neuron when a neurotransmitter or neuromodulator molecule binds to a G-protein-linked receptor, the G-protein dissociates into an α and a β-γ subunit. The latter directs a specific kinase to phosphorylate specific serines and threonines on the carboxy terminal of the receptor protein (Carman and Benovic 1998; Laporte et al. 1999). These kinases come in six types and subfamilies, each regulated by a different Ca^{2+} sensor protein (Iacovelli et al. 1999). This results in a conformational change in the receptor protein and increases its affinity for another membrane protein, beta-arrestin, in conjunction with N-ethylamide-sensitive fusion protein (NSF) (McDonald et al. 1999). This causes the receptor protein to dissociate from the G-protein and directs the former to a nearby clathrin-coated pit. During this process, the receptor becomes desensitized for its own transmitter.

In many cells clathrin-coated pits cover 2% of the cell surface (Schwartz 1995) and contain the remarkable protein, clathrin. This has three heavy chains that form a triskelon (figure 2.1) and three loose light chains attached to the ends. The molecules of clathrin self-associate to form a frame, like a BuckminsterFuller dome, composed of hexagonal and pentagonal units (Mukherjee et al. 1997; for details of the molecular biology of clathrin, see Pearse et al. 2000). The pit deepens to form a flask and its neck thins. A molecule of another protein, dynamin, a large GTPase, then wraps itself around the neck to pinch it off to form a vesicle. It has been suggested that dynamin may not be the final scissor itself, but works in association with an effector, endophilin, which is highly enriched in nerve terminals and which forms a complex with dynamin and synaptojanin (Ringstad 1999; Stenmark 2000).

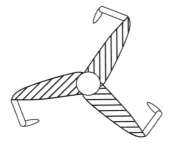

Figure 2.1
Clathrin—a triskelon protein.

This complex may form a functional link between dynamin and membrane phospholipid signaling, since synaptojanin is a polyphosphoinositide phosphatase and endophilin is a lysophosphatidic aryl transferase (Brodin et al. 2000). Amphiphysin also forms a stable complex with dynamin and synaptojanin, which excludes endophilin (Micheva et al. 1997). Endophilins also form a possibly nonfunctional complex with amphiphysin and synaptojanin, but at distinct sites (Cestra et al. 1999). Thus there are two established functional complexes involving these molecules: (1) endophilin-dynamin-synaptojanin and (2) amphysin-dynamin-synaptojanin (R. de Wit, personal communication).

Amphiphysins are closely connected with clathrin-mediated endocytosis. It has been proposed that amphiphysins drive the recruitment of dynamin to clathrin-coated pits (Wigge and McMahon 1998; Owen et al. 1998). Recently an alternative scheme has been suggested (Hill et al. 2001) in which dynamin is required for the late stages of invagination of the clathrin-coated pit (requiring hydrolysis of GTP), but that the Src-homology region 3 (SH3) domain of endophilin inhibits both the late stage of invagination and the scissoring. This latter effect is due to the lowering of phosphatidylinositol-4,5-biphosphate levels which in turn leads to dissociation of clathrin adaptor protein-2 (AP-2), clathrin, and dynamin from the plasma membrane.

Dephosphorylation of amphiphysins promotes complex formation between dynamin-1, synaptojanin-1, clathrin, AP-2, and amphiphysin, all of which are components of the endocytotic machinery. On the other hand, phosphorylation of dynamin-1 and synaptojanin-1 inhibits their binding to amphiphysin, and phosphorylation of amphiphysin inhibits its binding to AP-2 and clathrin. Thus phosphorylation regulates the association and dissociation cycle of the clathrin-based endocytotic machinery. Furthermore, calcium-dependent dephos-

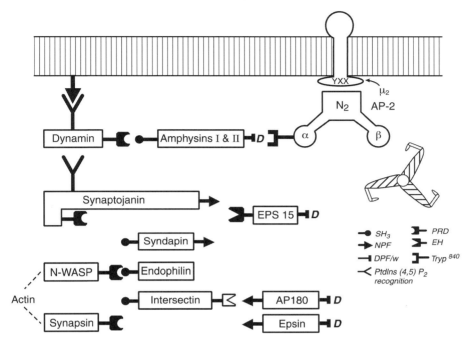

Figure 2.2
Some proteins involved in endocytosis: SH3 (Src-homology region-3 domain), NPF (arginine/proline/phenylalanine domain), DPF/W (arginine/proline/phenylalanine or tryptophan domain), PRD (proline-rich domain), and EH (Eps15-homology domain).

phorylation of endocytotic proteins could prepare nerve terminals for a burst of endocytosis (Slepnev et al. 1998). Simpson et al. (1999) also report, from in vitro experiments, that the SH3 domains of intersectin, endophilin I, syndapin I, and ampiphysin II inhibit clathrin-coated vesicle formation through interactions with membrane-associated proteins. The complex signaling system mediated by these proteins is shown in figure 2.2).

Koslov (1999) has suggested that dynamin molecules bind to the lipid membrane and self-assemble to form a helicoid structure. As a result of the hydrolysis of GTP induced by the GTPase activity of dynamin, this helix undergoes a change in its pitch that results in a dramatic change in the tubular lining of the neck of the developing vesicle constricted inside the helix, which serves to pinch it off. A review of other ingenious mechanistic hypotheses for the "pinchase" activity of dynamin has been presented by McNiven et al. (2000). These authors also present evidence that dynamins (for which 25 different mRNAs are known) have functions in addition to pinching off clathrin-coated vesicles. These include pinching off caveolin-based vesicles, phagocytosis, endosomal trafficking, actin function, and mitomorphology. Dynamin also has a functional role in mitogenic signaling, which is based on a different mechanism from its role in endocytosis (Whistler and von Zastrow 1999).

The protein mHip1R, a close relative of huntingtin-interacting protein (HiP), is another component of clathrin-coated pits and vesicles and may link the endocytotic machinery to the actin of the cytoskeleton (Engqvist-Goldstein et al. 1999). GIT1 is a GTPase activating protein that is involved in many instances of regulation of clathrin-mediated endocytosis that use beta-arrestin and dynamin (Claing et al. 2000). In general, arrestins bind G-protein-coupled receptors after their phosphorylation by a G-protein coupled (GR) kinase, which terminates agonist-mediated signaling. The phosphorylated receptor is capable of only minimal desensitization until it binds an arrestin protein and so engages the endocytic system (Leof 2000).

Endophilin is an enzyme that converts arachidonic acid plus lysophosphatidic acid (A) to phosphatidic acid (B). This serves to induce the alteration of membrane curvature necessary to form the vesicle (Murthy 1999). A and B are pyramidal-shaped molecules that insert into the membrane, as shown in figure 2.3. Appropriate packing of the membrane by these pyramids results in changes in the membrane shape needed for scissoring.

The new vesicle containing the endocytosed receptor is trafficked to the early endosome. The endosome system consists of a series of interconnected tubes and vesicles that are very actively mobile (Clague 1998). The inside of the clathrin vesicle is lined with a packing protein,

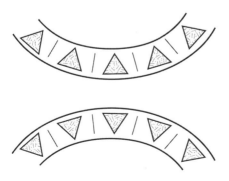

Figure 2.3
Phosphatidic membrane curvature molecules.

the clathrin adaptor protein (Clague 1998; Zhang et al. 1999), which fills the gap between the clathrin skeleton and the cargo molecule(s). There are several types of clathrin adaptor protein, two of which (AP-2 and AP-180) are involved in the endocytosis of receptors and synaptic vesicles (Zhang et al. 1999). These require phosphoinositides for their action (Gaidarov and Keen 1999).

Ligand-induced clathrin-mediated endocytosis is very fast and 2000–3000 vesicles per minute can bud off from a cell's external membrane in culture (Bottomley et al. 1999). Constitutive receptor-mediated endocytosis, which is also clathrin mediated, is slower (R. de Wit, personal communication). In some cases, in particular, certain neuropeptides, it is known that the ligand is endocytosed together with its receptor. The polypeptide is trafficked to the nucleus, where it modulates transcription (Jans and Hassan 1998). In the case of other smaller molecules such as dopamine or serotonin, it is not known whether endocytosis of the ligand occurs. However, since the inhibitor constant for the binding of dopamine is 0.9 nM, which indicates tight binding, it is likely that dopamine is endocytosed together with its D1 receptor (DIR).

Not all endocytosis occurs via clathrin-coated pits. There are also clathrin-independent endocytotic mechanisms, as in the case of angiotensin and the muscarinic acetylcholine receptor (Vickery and von Zastrow, 1999). The endocytosis of the dopamine D1 receptor is clathrin dependent, whereas that of the dopamine D2 receptor (D2R) is not. Moreover, some proteins can cross membranes by micelle formation without endocytosis (Prochiantz 2000).

Upon arrival at the early endosome, the vesicle loses its clathrin coat by the action of Hsc 70 (an ATPase heatshock cognate protein that is 70 kD) together with auxilin and possibly synaptojanin (Brodin et al. 2000). Its membrane then fuses with the endosome membrane and delivers its load into the interior of the early endosome. This is very acid (pH ~3.5), which causes the receptor to dissociate from its cargo (e.g., a polypeptide, an NM, or iron). The pathway for the receptor protein divides into two streams at this point. Part is carried to the late endosome by special large vesicles—a process that requires microtubules; the protein motors, kinesin and dynein (Pfeffer 1999); and a microtubule-binding protein the cytoplasmic linker protein-170 (CLIP-170) (Valetti et al. 1999)—and thence to the lysosome. Here the protein is broken down into its individual amino acids, which are recycled. The rest is recycled back in other vesicles to the external membrane, where the recycled receptor joins the receptor pool at the synapse.[1]

The continuous membrane of the endosome nevertheless contains distinct domains occupied by different regulatory Rab proteins (Sönnichsen et al. 2000). Three major populations of these have been reported—one that contains only Rab5, a second with Rab4 and Rab5, and the third with Rab4 and Rab11.

Distinct G-protein-coupled receptors can differ significantly in endocytic membrane trafficking after endocytosis by the same mechanism. For example, the beta 2 adrenergic and the delta-opioid receptors are both endocytosed by the same clathrin-associated process, but they are differentially sorted in the endosome system. The former is largely recycled (and shows negligible agonist-induced downregulation), whereas the latter is largely trafficked to the lysosome system (and shows substantial agonist-induced downregulation).[2]

Much work has been carried out on the endocytosis of the iron-carrying protein, transferrin. Upon binding a molecule of iron, the transferrin receptor is rapidly endocytosed by a clathrin-coated vesicle and is almost entirely recycled rather than being trafficked to the liposome for degradation (Schwartz 1995). In the acidic early endosome, the iron and transferrin dissociate. Further carrier proteins for transmission take up the iron to the several iron-containing enzymes in the neuron. The transferrin, together with the transferrin receptor, is recycled back to the external membrane.

Many G-protein-linked receptors have been shown to be endocytosed (for a list, see appendix B). Glutamate AMPArs are constitutively endocytosed at a slow rate, but are rapidly endocytosed upon binding the transmitter molecule. This needs concomitant activation of NMDArs (Carroll et al. 1999; Lissin et al. 1999; Morales and Goda 1999).

The rate of constitutive dynamin-dependant endocytosis of AMPArs is accelerated by insulin (Man et al. 2000). Neurons synthesize and release insulin. Since glucose utilization in neurons is largely insulin independent, this suggests that insulin in neurons has some other functions. These are listed by Man et al. (2000) as growth, maturation, protection, neuromodulation, learning, and memory. Insulin recruits GABAr-A and other receptors to the postsynaptic domain (Man et al. 2000). In *Xenopus* oocytes it also promotes a rapid delivery of NMDArs to the cell surface mediated by SNARE-dependent exocytosis (Skeberdis et al. 2001).

AMPArs stimulated by AMPA are endocytosed and largely recycled whereas AMPArs stimulated by insulin are endocytosed and are not recycled but are subjected to nonlysosomal degradation (Sheng and Lee 2001). Tetanic stimulation of neurons results in rapid delivery of the GluR1 subunit of the AMPAR to spines. This process also requires concomitant NMDAr stimulation (Shi et al. 1999). It is interesting that the specific inhibitor of phosphatidylinositol-3 kinase (PI-3K), wortmannin, does not inhibit the endocytosis of AMPArs whereas wortmannin plus insulin does (Sheng 2000). Early in cell stress, AMPArs are degraded by the protease, caspase, presumably following endocytosis. The function of this may be to ensure apoptosis and prevent the far more damaging excitotoxic necrosis (Glazner et al. 2000). The further details of the mechanism by which activation of the AMPAr initiates endocytosis of the receptor are not known.

Further details of the endocytosis of glutamate receptors are reviewed by Haucke (2000), based on reports by Beattie et al. (2000a) and Lin et al. (2000). Activation of AMPA receptors triggers endocytosis of AMPArs by a process involving L-type Ca^{2+} channels. This involves slow recycling via late endosomes (Burrone and Murthy, 2001). Curiously, AMPAr blockade also leads to endocytosis of these receptors, but without activation of L-type Ca^{2+} channels. NMDAr activation leads to AMPAr endocytosis by another mechanism that involves the Ca^{2+}-calmodulin-dependent phosphatase, calcineurin, a known regulator of endocytosis. This involves rapid recycling via early endosomes (Burrone and Murthy, 2001). Insulin triggering of endocytosis of the AMPAr does not require a Ca^{2+} influx, but is dependent on clathrin and dynamin. This leads to a very diffuse distribution of AMPArs throughout the neuron, in contrast to the very punctate distribution resulting from AMPAr and Nomad activation.

Heynen et al. (2000) present the first direct evidence in intact brain that synaptic glutamate receptor trafficking is bidirectionally and reversibly modified by NMDA receptor-dependent plasticity and that changes in glutamate receptor protein levels accurately predict changes in synaptic strength. They also state that LTP is associated with delivery of AMPA receptor protein to the synapse, whereas LTD is associated with the removal of both AMPA and NMDA receptor proteins from the synapse.

Glutamate NMDArs are one of the few receptors shown not to be endocytosed, or at least to be endocytosed very slowly over a period of hours or days (Lissin et al. 1999). Another receptor that is not endocytosed following activation is the alpha$_{2A}$ adrenergic receptor (Pierce et al. 2000a), although the beta$_2$ adrenergic receptor is endocytosed. However, endocytosis of the beta$_2$ adrenergic receptor is not necessary for its downregulation (Jockers et al. 1999). Curiously, activation of the alpha$_{2A}$ receptor leads to endocytosis of the EGF receptor (EGFr) even though its own receptor remains in the membrane (Pierce et al. 2000a).

Maudsley et al. (2000) have taken this further and have shown that activation of the beta$_2$ adrenergic receptor induces dimerization of the EGFr, tyrosine autophosphorylation, and internalization of this transactivated EGFr. The details of this process are that, upon binding its ligand, the beta$_2$ adrenergic receptor activates src kinase. This leads to formation of a complex between the beta$_2$ adrenergic receptor and the transactivated EGFr. Subsequently, there is recruitment of the complex to the clathrin-coated pit, followed by internalization and ERK phosphorylation. Endocytosis of the G-protein-coupled receptor (GPCr) is not required for ERK activation (Maudsley et al. 2000).

The beta adrenergic receptor is rapidly endocytosed whereas the vasopressin receptor is slowly endocytosed, both by a clathrin-beta-arrestin mechanism. One difference between these two is that beta-arrestin dissociates from the beta adrenergic receptor at the plasma membrane, whereas in the case of vasopressin, the beta-arrestin molecule is endocytosed along with the vasopressin receptor. These properties are completely reversed if the C-terminal tails of the two receptors are interchanged (Oakley et al. 1999).[3]

It should be noted that in many cases endocytosis is not limited to activated receptors, but includes the entire membrane. In goldfish retinal bipolar cells, the membrane recycles every 22 seconds as estimated by Bretscher and Aguado-Velasco (1998), or every 2 minutes as estimated by Rouze and Schwartz (1998). At small synapses on hippocampal pyramidal neurons, the external membrane recycles every 10 minutes (Murthy 1999). The average lifetime of an individual receptor molecule is 1–2 days (Conti and Weinberg 1999). It has recently been found that a complete glutamate synapse can be assembled in only 1–2 hours after initial contact between the axon terminal and the spine (Friedman et al. 2000). I will return to the significance of this later.

Membrane tension (e.g., during mitosis) also modulates the rate of endocytosis. Increased membrane tension decreases endocytosis (Raucher and Sheetz 1999) and increased endocytosis decreases membrane tension. This serves as a useful negative feedback control, which may operate also in the case of dendrite and spine expansion. Membrane tension also directly modulates stretch-sensitive integral membrane proteins. These include ion channels (for Cl^-, K^+, and Ca^{2+}) and enzymes (e.g., PLA2 and PLC). It also includes the NMDAr, which is downregulated by compression and upregulated by stretching (Ko and McCulloch 2000).

Thus, in a growing spine the pressure on the membrane overlying polymerizing actin could activate NMDArs and other stretch-sensitive proteins and thus lead to postsynaptic cascades, which would favor spine growth. Reid and O'Neil (2000) state that membrane tension is a key factor in regulating endocytosis and recycling in kidney cells.

Osmomechanically stretching the basolateral membrane of proximal tubule cells activates a Ca^{2+} channel, which promotes the insertion of new membrane. Membrane retrieval by endocytosis occurs at the borders of the basolateral membrane, leading to an overall membrane turnover. There is also membrane transcytosis from basal to apical regions, which transports K^+ and Cl^- channels. Endocytosis also plays a prominent role in mitogenesis via protein–protein interactions and protein–lipid interactions (see di Fiore and Gill 1999 for details).

There is also a link between the molecular machinery for endocytosis and that for the transport of proteins and RNA from the nucleus to the cytoplasm (Doria et al. 1999). This involves protein–protein interactions, i.e., Eps15 homology domain interaction between Eps15 and Hrb. Eps15, together with two other proteins, Ese1 and Ese2, also participates as a scaffold with Epsin1 and Epsin2 in the molecular rearrangement of the clathrin coats that are required for coated-pit invagination and vesicle fusion (Chen et al. 1998; Sengar et al. 1999; Rosenthal et al. 1999). The endocytosed material in many cases acts as a signal to modulate protein synthesis. The construction of new synapses may be partly mediated by shifting of the membrane by the endosome system, but it is also likely to require the synthesis of new proteins, especially in learning.

2.2 Some Enzymes Involved in Endocytosis

Important enzymes concerned in endocytosis are the protein phosphokinases and phosphatases discussed earlier that control endocytosis as well as many other neuronal functions. In addition, there are other enzymes with different functions.

2.2.1 Phosphatidylinositol Kinases

These enzymes phosphorylate membrane lipids at various positions of the inositol ring to generate second messengers, including phosphatidyl-3,4,5-triphosphate. The enzyme phosphatidylinositol-3 kinase plays a key role in endocytosis, being involved in vesicle trafficking (Folli et al. 1997), membrane recycling (Jess et al. 1996), and neurite extension (Kimura et al. 1994). It also plays a role in many other cellular mechanisms such as cell shape, adhesion, motility, proliferation, growth, differentiation, and survival (Krugman and Welch 1998).

Shpetner et al. (1996) report, from studies of the platelet-derived growth factor (PDGF) receptor, that PI-3K regulates the structure of the endosome and is involved in sorting of receptors during endocytosis. Different PI kinases play complex roles in the endocytosis of different receptors. For example, 3,4,5 kinase is involved in insulin receptor endocytosis, 4,5 kinase in clathrin-coated vesicle formation, 3 kinase in endosome function, 3 and 3,5 kinases in multivesicular bodies, and 3,5 kinase in lysosomes (Corvera et al. 1999).

Brain-derived neurotropic factor (BDNF) activity is mediated by PI-3K (Dolcet et al. 1999). PI-3K is also involved in regulation of the cytoskeleton by binding to profilin (Bhargavi et al. 1998). Activation of the transcription factor NF-κB by lipopolysaccharides and phorbol esters is mediated by PI-3K, whereas this activation induced by hydrogen peroxide, ceramide, and tumor necrosis factor utilizes some other pathway (Manna and Aggarwal 2000).

2.2.2 Small GTPases

The numerous changes in molecular architecture that characterize endocytosis, and synaptic plasticity in general, need energy. This is supplied in large part by a group of small GTPases. Small GTPases of the Ras superfamily play important roles in endocytosis (Ellis and Mellor 2000). This superfamily has some eighty members and is subdivided into six main families: Ras, Ran, Rad, Rab, Rho, and Arf.

Of these, Arf is involved in vesicle budding; Rab with vesicle targeting, and Rho with regulation of both endocytosis and the actin cytoskeleton. Rac and RhoA block clathrin-mediated endocytosis. In the developing retina, Rac1 and RhoA modulate the actin-dependent growth of terminal processes, which is also modulated by afferent glutamatergic transmission (Wong et al. 2000). Wong et al. suggest that these rapid movements may promote synaptogenesis. Endocytosed vesicles have to negotiate the band of polymerized actin that lies directly under the plasma membrane. Minor actin depolymerization leads to an increased rate of clathrin vesicle formation whereas major actin depolymerization has the opposite effect (Ellis and Mellor 2000). The Rho family regulate interactions between the intracellular tails of integrins and the cytoskeleton. This modulates the avidity of integrin molecules for extracellular matrix ligands (Schwartz and Shattil 2000).

Activated Rac and RhoA bind to caveolae, which are specialized rafts of concentrated glycosphingolipids and cholesterol coated with the protein caveolin, which are associated with actin. The role of caveolae in endocytosis is currently unclear except they seem to be involved in cholesterol transport in association with the hedgehog (Hh) signal transduction pathway (Incardona and Eaton 2000). The Hh receptor is a complex of two proteins, "patched" and "smoothener," plus cholesterol.

RhoD acts in early and/or recycling endosomes by regulating the traffic of vesicles along cytoskeletal tracks propelled by motor proteins. Rab5 controls the rate of endosome fusion. Rab11 is most likely localized in recycling endosomes, and Rab4 is present in all endosome preparations (Trischler et al. 1999). RhoB is involved in sorting multivesicular vesicles, where it phosphorylates some as-yet unidentified endosomal protein. One effect of this is to retard the rate of passage of the

EGF receptor through the endosome. Since EGF in its receptor is active even when endocytosed, timing its passage through the endosome in this way upregulates its effect. The mitogenic signals from the endocytosed EGF receptor end only when the lysosome is reached and the protein is degraded. EGF activation strongly increases Rho expression within 30–60 minutes. Another small GTPase, Ral, is also involved in EGF-mediated endocytosis. When EGF activates its receptor, a complex of Ral, Ral binding protein (RalBP1) and Partner for RalBP1, (POB1) transmit the signal to epsin and Eps15 (Nakashima et al. 1999; Morinaka et al. 1999). EGF also activates the MAPK cascade (Abe and Saito 2000).

Rho also organizes type 2 phagocytosis (in which the ligand sinks into an actin-lined invagination of the plasma membrane). Rac (and Cdc42) organize type 1 endocytosis (in which protrusions of the plasma membrane extend to engulf the object and to drag it into the cell) (Ellis and Mellor 2000).

Details of the mechanism by which Rab plays a key role in vesicle transport are as follows: Nascent vesicles recruit Rab + guanosine diphosphate, GDP, which changes to Rab + GTP during vesicle budding. The GTP molecule is trafficked together with Rab as it translocates. GTP hydrolysis may function as a switch that triggers the subsequent vesicle fusion with the target membrane and recycling of Rab (Vancura and Jay 2000).

Ras downregulates NMDAr phosphorylation and thus inhibits LTP (Manaabe et al. 2000).

2.3 Exocytosis

Until recently, exocytosis received much more attention from neuroscientists than did endocytosis because of its prominent role in the very visible function of neurotransmitter release. A great deal is now known about the mechanisms concerned. The exocytosis of synaptic vesicles in the presynaptic terminal involves the following sequence: The vesicle buds off the endosome and is trafficked to the presynaptic membrane. Here it undergoes a complex series of events: tethering (mediated by Rab), priming (mediated by NSF and SNAP), decking [involving SNARE complex formation between vesicle-associated membrane protein (VAMP), SNAP-25, and syntaxin], maturation, and finally fusion and delivery of its cargo into the synaptic cleft. The empty vesicle is then recycled to the endosome (Schiavo et al. 2000).

This mechanism consists of a complex series of protein–protein interactions According to the SNARE hypothesis, the core of these interactions is a heterotrimeric complex formed by SNAP-25, syntaxin, and VAMP-synaptobrevin (Cao and Barlowe 2000). Syntaxin also regulates chloride channels, and the GABA and glycine transporters (Geerlings et al. 2000). There is evidence that SNAP-25, but not synaptobrevin, is involved in constitutive axonal growth and dendrite formation (Grosse et al. 1999).

Other proteins interacting with this SNARE core are voltage-activated calcium channels, synapsins, annexins, and synaptotagmin (a putative calcium sensor). These are essential for the role of calcium in neurotransmitter release and synaptic plasticity (Verona et al. 2000).

Synapsins link vesicle phospholipids (by their globular heads) to proteins in the vesicle membrane (by their tails) as well as to the cytoskeleton. They are phosphorylated by cAMP-dependent protein kinase, calcium-calmodulin protein kinases, and proline-directed protein kinases (Damer and Creutz, 1994). Damer and Creutz suggest that phosphorylation of synapsins inhibits new actin filament growth and the binding of synapsin to synaptic vesicles and actin filaments, thus releasing the vesicle for exocytosis.

Annexins bind calcium and phosopholipids. They aggregate membrane vesicles and mediate membrane fusion. Synaptotagmin binds several proteins (e.g., neurexins and syntaxins), phospholipids, and calcium channels. It is involved in vesicle docking and membrane fusion (Damer and Creutz 1994). Synaptotagmin-IV is localized particularly in distal dendrites (Ibata et al. 2000). Leveque et al. (2000a) found that the calcium-triggered release of synaptotagmin precedes vesicle fusion. N-Ethylmaleimide-sensitive fusion protein may then dissociate ternary core complexes captured by endocytosis and recycle or prime individual SNARE proteins. Other proteins involved in this process include rabphilin-3A, rim, munc13, munc18, tomosyn, shapin, and NSF (see Mochida 2000 for details).

Exocytosis is followed by endocytotic membrane retrieval, which compensates for the increase in cell surface area. This retrieval is limited to the sites of previous exocytosis. This process requires the transitory insertion of calcium channels at the site (Smith et al. 2000).

2.4 Ubiquitation

An important role in endocytosis is played by ubiquitation. Ubiquitin is a small polypeptide that is covalently linked to receptor proteins by specific enzymes (Strous and Govers 1999; Kornitzer and Ciechanover 2000). Details of this remarkably complex process have been given by Levkowitz et al. (1999) in the case of the EGF receptor. When EGF binds to its receptor, a tyrosine kinase part of the EGFr molecule phosphorylates tyrosine-1045, which is located further toward the carboxy terminal end (figure 2.4a). This then attracts a molecule of Cbl adaptor

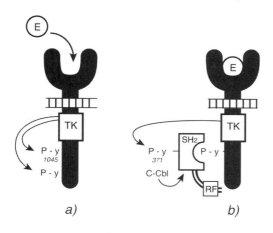

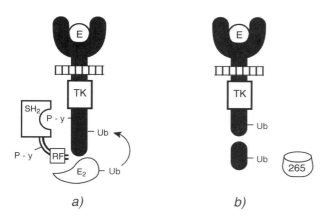

Figure 2.4
Mechanism of action of ubiquitin (see text). Y, tyrosine; P-y, phosphorylated tyrosine; C-Cbl, an adaptor protein; E, epidermal growth factor molecule; SH_2 and RF (ring finger) domains of C-Cbl; TK, tyrosine kinase; Ub, ubiquitin; and E_2, ubiquitin-activating enzyme.

protein, which has two domains—SH2 and RF. The SH2 domain binds to the now phosphorylated Tyr-1045 and the adjacent region of the EGFr molecule (figure 2.4b). When it does so, the tyrosine kinase on the EGFr phosphorylates Tyr-371 on the Cbl adaptor protein molecule. The EGFr-Cbl adaptor protein complex now attracts a molecule of the ubiquitin-activating enzyme E_2. This attaches a series of molecules of ubiquitin to sites on the EGFr molecule (figure 2.4c). This in turn attracts a proteosome (proteolytic enzyme), which fragments the intracellular portion of the EGFr (figure 2.4d).

These reactions normally take place in the cytoplasm and on the cytoplasmic face of the endoplasmic reticulum (Ciechanover et al. 2000). In the neuron, ubiquitation takes place during the transition of the EGFr molecule from the early endosome to the late endosome and these play a crucial role in sorting the endocytosed receptor to lysosomes as well as to endosomes (Lemmon and Traub 2000) for degradation.

Substrate recognition is mediated by a large family of ubiquitin ligases that bind different specific proteins to the ubiquitizing enzyme (Ciechanover et al. 2000). Ubiquitation tags damaged proteins and short-lived regulatory proteins for proteosomal degradation (Larsen et al. 1998; Rockwell et al. 2000). At the termination of this process, the ubiquitin is removed by specific deubiquitinylating enzymes (DUBs), of which four subtypes have been identified (Kawakami et al. 1999).

Ubiquitin is not only involved in recognition of proteins by proteosomes, it is also involved in downregulation of membrane receptors, transporters, and channels by promoting endocytosis. A member of the ubiquitin-protein ligase family, Nedd4/Rsp5p, plays a key role in this (Rotin et al. 2000). A number of ubiquitinlike proteins have recently been discovered; see Jentsch and Pyrowolakis (2000) for a review of this topic.

2.5 The Special Case of Endocytosis in Neurons

Neurons are very special cells and may be expected to express common biological mechanisms in idiosyncratic ways. This includes the role(s) of endocytosis.

In neurons, one important role of endocytosis relative to synaptic plasticity may be to redistribute membrane components to sites of new synapse formation (Hu et al. 1993). In this manner the endosome would act as a type of membrane bank for new synapse and spine formation. Ethell et al. (2000) have described the role of syndecan-2 and synbindin in promoting this transfer of membrane from the memory bank of the spine apparatus to the growing synapse. Dendrites and neuronal cell bodies have extensive networks of tubular early endosomes. There is a high rate of endocytosis in somatodendrites (Lewis and Lentz 1999).

In axons, endosomes only occur in axon terminals and boutons-en-passage. Late endosomes and lysosomes are predominantly localized to the soma. Endosomes inside neurons are also highly mobile and can be seen moving about in vivo by the use of special microscopic techniques (Parton et al. 1992). Spines also contain coated vesicles and multivesicular bodies, which indicates that endocytosis also occurs in spines (Spacek and Harris 1997).

Turrigano and Nelson (1998) have introduced the concept of "synaptic scaling" in which the rate and ratio of AMPAr removal from the subsynaptic membrane by endocytosis and its reinsertion by recycling may globally regulate synaptic strength at all of a neuron's synapses, or many of them. This concept can be applied to other receptors besides the AMAPr. Thus the growth of spines and synapses and their biochemical constitution in terms of the relative presence or absence of specific receptors and other related key membrane elements is a dynamic process subject to modulation of the ratio between endocytosis and recycling (Smythies 2000).

2.6 The Functional Significance of Membrane and Receptor Endocytosis

There are several putative functions of the endocytosis of membranes and receptors that apply to neurons:

- Desensitization of the receptor
- In some cases, removal of phosphate groups, which is necessary to resensitize the receptor
- Repair of oxidatively damaged proteins
- Capture by the cell of external molecules needed for specific purposes by the postsynaptic cell

2.6.1 Desensitization of Receptors

When a shower of neurotransmitter or neuromodulator molecules arrives at the postsynaptic membrane and binds to their receptors, the initial desensitization is mediated by phosphorylation of the receptor molecules (Ferguson and Caron 1998). Supportive evidence for this has been obtained by Jockers et al. (1999), who showed that in the case of the beta$_2$ adrenergic receptor, endocytosis is not a prerequisite for downregulation. The rapid internalization of these phosphorylated receptors will then maintain the desensitization simply by physical removal of the receptors (Ferguson and Caron 1998). The synapse will return to normal when these phosphorylated and endocytosed receptors are dephosphorylated and recycled to the external membrane.

On the other hand, Ogimoto et al. (2000) report the results of a study of the inhibition of Na^+, K^+-ATPase (NKA) activity by dopamine. This is normally mediated by phosphorylation of the receptor molecule and its subsequent endocytosis. However, a dopamine antagonist such as oxymetazoline, which interferes with the ability of dopamine to recruit the clathrin necessary for endocytosis, can reverse this inhibition. These authors conclude that inhibition of NKA activity can be reversed by preventing its endocytosis without altering the state of phosphorylation of its alpha subunit.

In some cases endocytosis is required for receptor desensitization only (e.g., secretin and concanavalin A, Con A receptors) and in other cases for resensitization only (e.g., adenosine A2 receptors) (Mundell and Kelly 1998). The process of endocytosis and exocytosis (recycling) occurs in quantum amounts in which the most frequent size of the quantum is $\sim$0.04 µm, corresponding to vesicles of about 110 nm in diameter (Henkel et al. 2000). The authors suggest that this process can act as a chemical oscillator. The role of arrestins in receptor desensitization (Leof 2000) was reviewed earlier.

The time needed for receptor recycling has been measured. Beta$_2$ adrenergic receptors take 10 minutes to reach the endosome (Cao et al. 1998). Dopamine D1 receptors take 4 minutes (Dumartin et al. 1998). The rabies virus takes 5 minutes (Lewis and Lentz 1999). The NGF receptor (NGFr) requires 30 minutes for the complete cycle (Zapf-Colby and Olefsky 1998). Each transferrin receptor recycles $\sim$300 times (Zapf-Colby and Olefsky 1998). A typical membrane protein with a half-life of 10–20 hours will make at least ten roundtrips to the endosome per lifetime (Kelly 1999).

2.6.2 Phosphorylation and Dephosphorylation

The phosphorylation of G-protein-coupled receptors is mediated by two classes of protein kinases: second messenger-dependent protein kinases (e.g., cAMP-dependent protein kinase, PKA) and G-protein-coupled receptor kinases (GrKs) (Ferguson and Caron 1998). The former have the capacity indiscriminately to phosphorylate and desensitize receptors that have not been exposed to the agonist, but they also fail to promote the binding of arrestins to the G-protein.

In contrast, GrKs specifically phosphorylate the agonist-activated form of the GPCr and promote the binding to it of the arrestins, which further uncouple the receptors by inhibiting the binding of G-proteins to the receptor protein. Thus GrKs diminish the cell's response only to the agonist to which it was exposed. However, there is much cross-talk between the two systems (Ferguson and Caron 1998).

In the case of the GluR1 subunit of the AMPAr, serine-831 can be phosphorylated (by alpha Ca^{2+}-calmodulin-dependent kinase II, CaMKII) and dephosphorylated (by PSD-associated protein phosphatase I, PPI). In contrast, serine-845 (phosphorylated by PKA) cannot be either phosphorylated or dephosphorylated in situ, but only when the AMPAr is no longer anchored to the PSD (Vinade and Dosemeci 2000). Furthermore, when dopamine activates its D1 receptor, this effects phosphorylation of the AMPA Glur1 subunit (via PKC) at the Ser-845 site, but not at the Ser-831 site (Snyder et al. 2000).

The phosphate groups attached to receptors by protein kinases following receptor activation constitute, as we have seen, the primary mechanism of receptor desensitization. In many cases these phosphate groups cannot be removed in situ in the membrane and can be removed only by the endosome following endocytosis of the receptor (Ferguson and Caron 1998). As these authors say, "The mechanism underlying GPCR resensitization is thought to be the agonist-dependent sequestration (endocytosis) of GPCRs to the endosomal compartment of the cell where they are dephosphorylated and subsequently recycled back to the plasma membrane as fully functional receptors" (Ferguson and Caron 1998, p. 122).

However, the possibility that in other cases dephosphorylation can be effected in situ in the membrane is suggested by the report that the A-kinase-anchoring protein (AKAP) has binding sites for two phosphokinases (PKA and PKC) and for one phosphatase, calcineurin (Sik et al. 2000). Thus AKAP brings into one locus enzymes for putting on phosphate groups and for taking them off again. AKAP is located near to, but not actually in, the postsynaptic density of pyramidal cells in the CA1 region of the hippocampus and subicular and presubicular regions. It is absent from the CA2 and CA3 and dentate gyrus regions of the hippocampus. PKA modulates the phosphorylation of AMPArs whereas PKC modulates the phosphorylation of NMDArs (Sik et al. 2000). AKAP dephosphorylates NMDArs (Tong et al. 1995; Lieberman and Mody 1994). Furthermore, NMDAr stimulation also activates a local dephosphorylating phosphatase that acts on the cAMP response element-binding protein attached to the PSD-95 scaffold protein (Sala et al. 2000).

Chern (2000) gives a good review of adenylate cyclases in the central nervous system. In particular, two isozymes of ACs ACII and ACIV, are concerned in synaptic plasticity. ACs are linked to the receptor by a G-protein and indeed AC itself can act as a G-protein (Chern 2000). ACs are controlled by various kinases. ACs can act as biochemical coincidence detectors, as in the case of Ca^{2+}-calmodulin-dependent kinase and 5-hydroxytryptamine-(5HT)-sensitive ACs, which are jointly involved in learning the withdrawal reflex in *Aplysia* (Chern 2000). The late phase of LTP requires activation of AC and cAMP-dependent protein kinase. The mechanism that mediates this effect has been described by Chain et al. (2000). Ubiquitin-mediated proteolytic degradation of the regulatory receptor subunit of PKA releases the catalytic subunit, which migrates to the nucleus to phosphorylate CREB. This results in upregulation of the protein synthesis needed to build new synapses. PKA is also important in early LTP at a postsynaptic locus of action (Otmakhova et al. 2000). Both PKA and MAPK are necessary for expression of a persistent phase of LTP that is dependent on protein synthesis in the lateral amygdala and is thought to be involved in fear conditioning (Huang et al. 2000).

A recent general review of the adenyl cyclase-dependent signaling pathways has been given by Fimia and Sassone-Corsi (2001).

2.6.3 *Repair of Oxidative Damage*

Receptors and other membrane proteins and lipids are under constant oxidative attack by ambient reactive oxygen species such as hydrogen peroxide, superoxide anions, and the hydroxyl radical, which are produced during the course of many enzymatic reactions in the cell. In order for the cell to survive, this damage must be repaired. There is no known mechanism to do this in situ in the membrane. The ubiquitin-protease pathway described earlier is required for this processing of oxidized proteins (Figueiro-Pereiro and Cohen 1999; Kornitzer and Ciechanover 2000; Grune and Davies 1997). The proteins to be repaired are endocytosed and trafficked to the endosome system. Here it is possible that the oxidized portions are ubiquinated and trafficked to the liposome for degradation. The intact proteins are recycled. Thus the endosome system may perform what I have called a "triage function" (Smythies 1997, 2000), which is familiar to military surgeons.

2.6.4 *Molecular Capture*

The fate of the ligand is known in the case of many polypeptide NMs and hormones. These are endocytosed inside their receptors and are trafficked via the endosome system to the nucleus to modulate transcription (Koenig and Edwardson 1997). As these authors say, "the purpose of endocytosis is to capture the ligand for subsequent use by the cell" (Koenig and Edwardson 1997, p. 282).

An example is provided by nerve growth factor. This is secreted by the target neuron for a growing axon. Upon binding to its receptor TrkA on the growing axon, it is rapidly endocytosed inside its receptor and trafficked to the nucleus, where it modulates transcription by autophosphorylation of tyrosine kinases (Grimes et al. 1996). The axon, now that it has contacted its target, needs a revised program of protein synthesis that is different from when it was growing and looking for its partner dendrite. The TrkA molecule may still exert an active signaling function while it is inside the endocytotic vesicle (Beattie et al. 1996; Skarpen et al. 1998). Recently it has been reported that NGF (as well as BDNF), upon binding to its receptor TrkA, increases the level of clathrin in the plasma membrane and stimulates endocytosis (Beattie et al. 2000b). In Alzheimer's disease, there is a profound fall in the number of trkA-containing neurons in the nucleus basalis (Mufson et al. 2000).

Not all transport of proteins across membranes is mediated by endocytic mechanisms. In the endoplasmic reticulum, there are preprotein translocases that are multimeric protein complexes that catalyze the transport of newly synthesized membranes across the membrane (van Voorst and de Kruiff 2000). Proteins can also cross membranes by micelle formation (Prochiantz 2000).

Chapter 3
Special Proteins

3.1 The Role of Cell Adhesion Molecules

The functional aspects of neurons related to neural plasticity are not confined to receptors and postsynaptic cascades. The mechanisms by which axon terminals and spines stick to each other in the appropriate pairing are also important. If a spine is pruned, its connection with its axon terminal must be dissolved. When a new spine replaces a pruned spine, the correct axon terminal must contact it. Most new spines contact terminals that already have other synapses on them (Harris 1999). One of the molecular bases for this is a series of cell adhesion molecules.

The main CAMs at the synapse include cadherins, neurexins, neurolignins, integrins, and members of the immunoglobulin superfamily such as transiently expressed axonal surface glycoprotein-1 (TAG-1)-contactin and members of the L1 family (e.g., neurofascins and Nr-CAMs). The general pattern for CAMs is that one transmembrane molecule has its N-terminal loop(s) in the synaptic cleft and its C-terminal loop(s) inside the neuron. The N-terminal loops bind to a complementary molecule (either homotropic or heterotropic) from the other side of the synaptic cleft. The C-terminal contacts the cytoskeleton either directly or via intermediary associated molecules (figure 3.1). Associated molecules include catenins, selectins (in blood cells), ankyrins, and GIT1 (Hagler and Goda 1998; Conti and Weinberg 1999; Hortsch 2000; Hynes 1999). When an axon is growing, its cytoskeleton needs to grow too. When the axon terminal contacts its appropriate spine, the CAM and associated molecules react by sending the signal "Stop growing" to the cytoskeleton. The cytoskeletal proteins such as actin then enter a new phase of organization. A similar process occurs in spines when they cease looking for a partner after firm binding by an axon terminal.

3.1.1 Cadherins
The cadherins from the two sides interact with each other by a zipper-mechanism in which one line of large lipophilic residues (such as

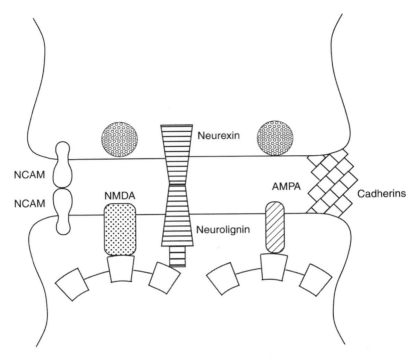

Figure 3.1
Some CAMs.

tryptophan, Trp) alternating with small lipophilic residues (such as alanine, Ala) interdigitate with a similar and complementary line from the other side (Hagler and Goda 1998). Le et al. (1999) present evidence that E-cadherins are subject to constant trafficking through the endosome recycling pathway, which they suggest may provide a mechanism for the availability of E-cadherin for junction formation in development and tissue remodeling. Antibodies against cadherin and a protocadherin, arcadlin (Shiosaka and Yoshida 2000) inhibit LTP.

3.1.2 Catenins and Related Molecules

Catenins mediate the interaction of cadherins with the underlying cytoskeleton and also act as signaling molecules to the nucleus acting on the Tcf/Lef-1 family of transcription factors (Murase and Schuman 1999; Giannini et al. 2000) as part of the "wingless" (Wnt) signaling pathway (Kikuchi 2000) (figure 3.2). Wnts and Hedgehog [called "sonic Hedgehog" in vertebrates (Britto et al. 2000)] are morphogens that act

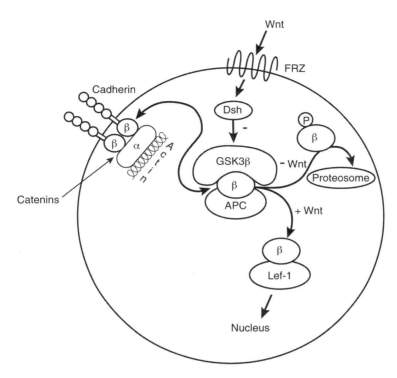

Figure 3.2
Mechanism of action of glycogen synthase kinase (GSK) and Wnt. Lef-1, transcription factor; β, beta catenin; Dsh, dishevelled; APC, adenomatous polyposis coli; FRZ, frizzled; P, phosphorus.

in embryogenesis to specify different cells by activating the transcription of different genes (Christian 2000).

In later life, postsynaptic cerebellar granular cells secrete glycoproteins, including Wnts. They bind to "frizzled" receptors on the growth cones of presynaptic mossy fibers and stabilize beta-catenin (Behrens 2000). The beta-catenin is trafficked to the nucleus and modulates transcription of various genes (e.g., c-myc, c-Jun, cyclin D1, fra-1). In the absence of Wnts, the beta-catenin becomes unstable, is degraded by a kinase glycogen synthase kinase-3β (GSK-3β), and is cleared by ubiquitation and proteosomes. The scaffolding proteins for this operation are axin plus conductin (Behrens 2000; Kikuchi 2000). This results in remodeling of the growth cone (cone size and complexity and axon thickness and spreading) (Hall et al. 2000). Wnt-7a also induces synapsin-1 clustering, which is a preliminary step in synaptogenesis

(Hall et al. 2000). It has been suggested (Salinas 1999) that Wnt-7a induces axonal remodeling by inhibiting glycogen synthase kinase-3β. Inhibition of this enzyme leads to a decrease in a phosphorylated form of microtubule-associated protein-1B (MAP-1B, a protein involved in microtubule assembly) and thus to a concomitant decrease in the level of stable microtubules. There is considerable cross-talk between the Wnt and MAPK pathways (Behrens 2000).[1]

3.1.3 The Immunoglobulin Superfamily and Related Molecules

The immunoglobulin (I_2) superfamily molecules (such as neurolignins, nectins, neurexins, and members of the L1 family, which includes L1-CAMs, Nr-CAMs, and neurofascins, and contactin) are also located on each side of the synapse. Nr-CAM molecules from the presynaptic and postsynaptic sides make homotropic contacts. Nectins are members of the immunoglobulin superfamily of adhesion molecules located at cadherin-based adhesion sites. They connect via afadin to the actin cytoskeleton (Satoh-Horikawa et al. 2000). Beta-neurexin from the presynaptic side binds by a heterotropic mechanism to neurolignin from the postsynaptic side (Conti and Weinberg 1999). Nr-CAM can bind by this homophilic mechanism, but it also functions as a neuronal receptor for neurite growth-promoting stimuli provided by F11 contactin and neurofascin. It can also interact laterally within the external membrane with F11 contactin in the form of a complex that may transmit signals to regulate axonal growth (Grumet 1997). Nr-CAM-stimulated neurite growth is mediated by two separate pathways—the Ras-MAPK pathway and the cascade fibroblast growth factor (FGF) receptor → PLC → PKC (Kolkova et al. 2000).

Mitogen-activated protein kinase[2] is required for LTP in the hippocampus. MAPK is activated in the hippocampus during training and is necessary for contextual fear conditioning (Walz et al. 2000). Ephrin-A and its receptors acting upon PDZ[3] scaffolding proteins, and thus affecting actin polymerization (Gerlai and McNamara 2000), mediate memory of fear conditioning. Recent evidence (Szapiro et al. 2000) suggests that retrieval of long-term memories depends on mGlur and MAPK activity whereas storage of such memories depends rather on NMDAr and calcium-calmodulin-dependent protein kinase II activity.

The major Nr-CAM isoform 180 is found in some synapses, but not in others (Wheal et al. 1998). During active avoidance training, polysialic acid containing forms of Nr-CAMs redistributes to the edges of the active zone of the synapse. Thus learning is associated with the increased expression and redistribution to this Nr-CAM isoform. LTP is inhibited by antibodies against neural cell adhesion molecules (NCAM) and L1 (Shiosaka and Yoshida 2000).

Neurofascin is a member of the L1 subgroup of the Ig superfamily that promotes axon growth by interactions with NCAM. It also interacts with the extracellular matrix glycoprotein, tenascin R, and the Ig superfamily members axonin-1 and F11 (Volkmer et al. 1998). Tenascin R and tenascin C also bind to the axon-associated protein F11 in a complex manner. These two forms of tenascin act in different ways. Tenascin R increases cell attachment and neurite growth whereas tenascin C increases cell attachment but decreases neurite growth (Zacharias et al. 1999). Both forms of tenascin are able to form larger molecular complexes and to link F11 polypeptides by forming a molecular bridge. Tenascin C may also be involved in synapse elimination (Wheal et al. 1998). Furthermore, antitenascin antibodies injected into the brain of a newly hatched chick totally inhibit acoustic imprinting. Thus tenascin would appear also to mediate this form of learning (Wheal et al. 1998).

Neurofascin is also required to maintain physiological levels of voltage-gated sodium channel activity (Zhou et al. 1998). L1-mediated axon growth is mediated by activation of the MAPK cascade and depends on PI-3K activity and internalization of the L1 molecule (Schaefer et al. 1999; Schmid et al. 2000).

The L1 molecule is endocytosed at the stem or C region of the growing axon. It is then trafficked in vesicles along microtubules to the growing tip (growth cone), where it is cycled to the surface. This serves to pull the growth cone along (Kamiguchi and Lemmon 2000). Expression of L1 is upregulated in regenerating cholinergic axons during axonal regeneration and is downregulated upon target contact (Aubert et al. 1998). In humans, L1 mutations cause the clinical syndrome of mental deficiency called CRASH (corpus callosum agenesis, retardation, adducted thumbs, spastic pariesis, hydrocephalus) (Schmid et al. 2000).

Mint-1 and mint-2 are brain-specific proteins that play a role in the machinery for synaptic vesicle fusion (Okamoto and Südhof 1998) in a complex with the alpha$_{1B}$ subunit of the voltage-gated calcium channel and CASK (a membrane modular adapter protein kinase that contains the signaling modules CAM kinase and SH3) (Maximov et al. 1999).

In myelinated fibers, the cell adhesion molecules neurofascin and NCAM are linked to the cytoskeleton by ankyrin (Scherer 1999). The membrane-skeletal protein, ankyrin G, is essential for the proper clustering of voltage-gated sodium channels (Zhou et al. 1998).

3.1.4 Integrins and Related Molecules

Integrins from one side of the synapse interact on the other side of the synapse with another CAM, Ig-SF (immunoglobulin superfamily). Integrins are linked to actin via the actin-binding proteins, talin,

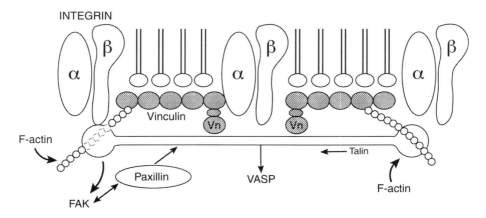

Figure 3.3
Vinculin and related molecules. Vh, vinculin head; vt, vinculin tail; FAK, focal adhesion kinase; VASP, vasodilator-stimulated phosphoprotein.

alpha-actinin, and filamin, as well to the cytoskeletal protein, vinculin (Critchley et al. 1999). These actin-binding proteins not only couple integrins to the cytoskeleton but also provide a surface for the attachment of a number of other signaling molecules. Talin possesses binding sites for the nonreceptor protein tyrosine kinase, focal adhesion kinase (FAK), layilin, and other membrane and cytoskeletal proteins. FAK and protein-rich tyrosine kinase2 (PYK2), are highly expressed in the brain.

Mu-opioid agonists specifically activate FAK by tyrosine phosphorylation (and also increase the phosphorylation of the cytoskeletal protein, cortactin, and of the focal adhesion site protein, vinculin) (Mangoura 2000). Mu-opioids also activate MAPK (in a manner independent of endocytosis of this receptor), inhibit adenylate cyclases, and modulate calcium channels (Trapaidze et al. 2000). FAK and PYK2/CAKbeta are differentially regulated by neurotransmitters and depolarization and have different but overlapping distributions (see Menegon et al. 1999 for details).

Vinculin, besides binding to talin, alpha-actinin, and actin, also binds to the cytoskeletal protein, paxillin. These interactions are summarized in figure 3.3 (Critchley et al. 1999). Paxillin is a substrate for FAK and is thought to act as a signal that functions to link other proteins into multimolecular signaling complexes (Stevens et al. 1996; Berg et al. 1997; Renaudin et al. 1999). The mode of action of vinculin is ingenious. In its inactive state, the tail of the molecule binds to its head and covers all the binding sites except that for paxillin. Upon activation by phosphoinositide P2, the molecule unwinds and this uncovers the other binding sites. The phosphoinositide P2 is synthesized by the small GTPase, Rho. Vinculin and debrin form separate microfilament (actin) anchoring plaques in many cells, including neurons in cell–cell and cell–matrix attachments (Peitsch et al. 2000).

Filamin binds to integrin. A mutation in the filamin gene results in the disease, periventricular heteropia. A newly discovered family of docking proteins involved in integrin signaling, Cas proteins, have recently been reported (O'Neill et al. 2000). These interact with FAK. They assist in transducing integrin signals to the nucleus and are involved in cell movement and changes in cell shape.

Talin acts on filopodial motility and may couple both filopodial extension and retraction to actin dynamics. Vinculin is not required for filopodial extension and retraction, but rather plays a role in the structural integrity of filopodia (Sydor et al. 1996).

The activity of integrin is also modulated in some instances by reelin. Reelin is a protein that is synthesized by GABAergic interneurons in layer I of the cerebral and cerebellar cortices and released into the extracellular space. Here it is taken up by apical dendritic spines, where it complexes with integrin in the PSD. The reelin-integrin pathway modulates the phosphorylation of disabled-1 (Dab1) protein in spines and so affects the cytoskeleton (Rodriguez et al. 2000). A recent review of the integrins is given by Humphries (2000). Reelin levels and the expression of reelin mRNAs are reduced in the cortex in schizophrenia (Guidotti et al. 2000).

3.1.5 Further Functions for CAMs
CAMs modulate both cell adhesion and neuritic growth. These functions are mutually antagonistic and are controlled by the level of expression of individual CAMs, changes in their molecular isoforms, post-translational modifications, and intracellular signaling (Fields and Itoh 1996). Neurotransmitters can regulate CAM expression, and CAMs are involved in synaptic plasticity and in the protein synthesis-dependent phase of learning.

Some CAMs not only keep axon terminals and spines in contact but also have further functions inside the postsynaptic neuron. For example, syndecan-2 is located on spines and promotes their growth. Its external glycosaminoglycan chains can bind extracellular matrix ligands, and its core protein cytoplasmic domain can signal during cell adhesion (Couchman and Woods, 1999). Syndecan clusters recruit the scaffolding protein, CASK, which plays a role in the organization of the cytoskeleton and the signaling molecules required for the development of the mature spine (Ethell and Yamaguchi 1999). Syndecan operates in conjunction with PKCα and phosphatidylinositol (Horowitz et al. 1999). Recently Ethell et al. (2000) have shown that syndecan-2, after binding to its receptor synbindin in the plasma membrane, induces spine formation by recruiting intracellular vesicles from the spine apparatus (a series of vesicles located under the PSD) to supply the new membrane and receptors need for spine construction.

Burden-Gulley and Lemmon (1996) grew retinal ganglionic neurons in culture and showed that the pattern of three cytoskeletal elements (F-actin, microtubules, and neurofilaments) differed in accordance with the presence or absence of the CAMs N-cadherin, L1, and laminin in the nutrient medium. They concluded that these CAMs directly influence the cytoskeletal morphology of the growth cone. Cotman et al. (1998) point out that amyloid precursor protein shares many of the properties of a classical CAM and that beta-amyloid can masquerade as a pseudo-CAM, a feature of relevance to Alzheimer's disease.

CAMs can also directly influence events in the postsynaptic neuron. For example, NCAM activates PLCγ to produce the second-messenger AA (Doherty et al. 1995). CAMs are also subject to endocytosis (Doherty et al. 1995). NCAM also acts as the receptor for the neuronal adhesion glycoprotein F3 in its inhibitory effect on axonal elongation (Faivre-Sarrailh et al. 1999). Neurolignins secreted by non-neuronal cells can induce morphological and functional presynaptic differentiation in contacting axons. Thus neurolignins are a part of the machinery employed during the formation and remodeling of synapses (Scheiffele et al. 2000).

3.1.6 Proteoglycans

Proteoglycans consist of a polypeptide core and a glycosaminoglycan moiety. They constitute a part of the extracellular matrix and modulate its organization. There are seven main types of proteoglycan (including hyaluronic acid, chondroitin sulfate, and heparin). They are involved in growth factor action, cellular adhesion, motility, and migration, as well as axonal and dendritic outgrowth. Growth promoters of this type can promote growth or inhibit it, depending on their molecular composition, how they are presented to neurons, and the type of neuron involved. Other growth promoters will be discussed later.

3.1.7 Some Chemical Mechanisms That Make and Break Cellular Contacts

To help effect synaptic deletion, NMDAr stimulation causes the postsynaptic cells to excrete a serine protease, which leads to the rapid extracellular proteolysis of cell adhesion molecules (Hoffman et al. 1998) and loosening of axon–spine contacts. Serine proteases modulate the early phase of LTP (Shiosaka and Yoshida 2000). Furthermore, low calcium levels in the synaptic cleft that results from NMDAr activation (when the calcium pours into the postsynaptic neuron) destabilize the molecular bonds that cause CAMs to stick to each other (Murase and Schuman 1999).

A third mechanism by which NMDAr activation can affect synaptic adhesion is described by Tanaka et al. (2000). This activation causes *N*-cadherin to dimerize and become more resistant to protease degradation and thus to fasten the axon terminal more firmly to its spine. Thus synaptic events can affect the cytoskeleton via direct molecular contacts. Furthermore, activation of catenin in the postsynaptic neuron could alter presynaptic mechanisms via their link through cadherins, thus lessening the need for retroactive transmitter molecules (Hagler and Goda 1998).

3.2 Scaffolding Proteins

A scaffolding protein holds other proteins, particularly receptors, enzymes, and elements of the cytoskeleton, in a close functional relationship to each other. Under the postsynaptic membrane there is a layer visible in an electron microscope picture that is called the "postsynaptic density." This is a highly organized structure that contains a variety of scaffolding proteins, such as postsynaptic density protein-95 (PSD-95), CASK, glutamate receptor-interactive protein (GRIP), myristo-lated alanine-rich C kinases (MARCKs), A-kinase-anchoring proteins (AKAPs) and membrane-associated guanylate kinase (MAGUKs), and at a deeper layer, Shank (a family of synaptic proteins) (Ehlers 1999; Fanning and Anderson 1999; Ramsden 2000; Naisbitt et al. 1999). On the external side, these proteins bind a variety of receptors and related enzymes, ion channels, and transsynaptic proteins (e.g., neurolignins and neurexins) in a precisely engineered order. On the internal side, scaffolding proteins link with other enzymes and with the actin of the cytoskeleton.

MAGUKs cluster glutamate receptors whereas AKAPs cluster kinases and phosphatases (Colledge et al. 2000). GRIPs are scaffolding proteins that tether AMPA receptors (Dong et al. 1999). NSF also stabilizes AMPAr clusters in the membrane, but it may do this by acting as a chaperone protein for the AMPAr molecule (Turrigiano 2000).

The membrane-associated MAGUK protein, guanyl kinase PSD-95, scaffolds NMDA receptors. It suppresses the endocytosis of the potassium channel protein Kvl.4. In transfected cells, Kvl.4 was endocytosed with a half-life of 87 minutes. This was completely suppressed in the presence of PSD-95 (Jugloff et al. 2000). The authors suggest that this finding represents a fundamentally new role for MAGUKS such as PSD-95 in the stabilization of ion channels at the cell surface. PSD-95 also binds to another protein, cyprin, which regulates postsynaptic protein sorting (Firestein et al. 1999). Direct interactions between the MAGUK proteins PSD-95 and (synapse-associated protein (SAP-97) on the one hand and AKAP-79/150 on the other, form a complex that recruits PKA and mediates phosphorylation of the AMPAr (Colledge et al. 2000).

As detailed earlier, neurexins are neuronal cell-surface proteins with up to thousands of isoforms. They bind tightly across the junction to a second class of neuronal cell-surface receptors, the neurolignins. Intracellularly, the neurexin–neurolignin junction is bound by CASK on the neurexin side and PSD-95 on the neurolignin side (Missler et al. 1998) via PDZ domains. CASK is a MAGUK required for EGF localization and signaling (Hsueh et al. 2000). In the adult rat brain, it is concentrated at synapses and binds to the cell surface proteins, neurexin and syndecan, and to the cytoplasmic proteins, mint and veli. It also binds, via its guanylate-cyclase domain to Tbr1 (T-box transcription factor), that is involved in forebrain development. The CASK–Tbr1 complex enters the cell nucleus and binds to a specific DNA sequence (the T-element). The genes activated by this interaction include reelin, which is essential for cortical development (Hsueh et al. 2000).

Shank is a recently discovered family of postsynaptic proteins that function as a major deeper-layer scaffolding protein linking AMPArs, NMDArs, and mGlurs (Tu et al. 1999; Naisbitt et al. 1999) (figure 3.4).

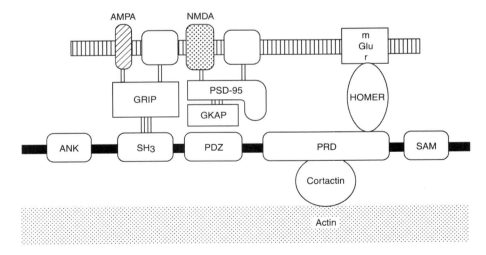

Figure 3.4
Shank and related proteins. Five domains of Shank are shown: ANK (ankyrin repeats 1–7); SH3 (Src homology 3), PDZ (PSD-95 domain), PRD (protein-rich domain), and SAM (sterile alpha motif). GKAP, guanylate kinase domain-associated protein; GRIP, glutamate receptor-interacting protein.

The Shank protein binds to GRIP from its SH3 domain (and thus anchors the AMPA receptor), to guanylate kinase domain-associated protein (GKAP) from its PDZ domain (and thus anchors the NMDAr via PSD-95), and to Homer from its main body (and thus anchors the mGlur). It also binds cortactin (and thus anchors itself to the actin cytoskeleton).

Guanine nucleotide exchange factors (GEFs) actively regulate the activity of the Ras, Rho, Rab, Ran, and Arf families of small G-proteins. GRASPs (GRIP-associated proteins) are neuronal Ras GEFs associated with GRIP and AMPArs. Their activity is regulated by NMDAr activity. The GRIP-associated protein, GRASP-1, may regulate neuronal Ras signaling and contribute to the regulation of the AMPAr redistribution that follows NMDAr activity (Ye et al. 2000).

Homer proteins form multivalent complexes that bind proline-rich motifs in group 1 mGlurs and inositol triphosphate receptors, thereby coupling these receptors to a signaling complex. Homer-1c increases the cell surface expression of the mGlur type 1 alpha by increasing its retention within the membrane (Ciruela et al. 2000). In the brain, Homer-1c colocalizes with mGlurs in the cerebellar molecular layer, but not in the hippocampus. In the latter, it colocalizes in dendritic spines with the NMDA receptor. Homer-1c acts to induce neurotransmitter receptor clustering. In this clustering, an array of leucines on the extreme C-terminal of the protein acts as a zipper (Tadokoro et al. 1999). Homer also couples group 1 metabotropic glutamate receptors to N-type calcium and M-type potassium channels (Kammermeier et al. 2000).

An isoform of Homer called "cupidin" (or Homer-2a) acts as a scaffolding protein, linking mGlurs with the cytoskeleton and Rho family proteins in cerebellar granule cells (Shiraishi et al. 1999). Another isoform of Homer (Homer-1b) inhibits the surface expression of mGlur 5, causing it to be retained in the endoplasmic reticulum. Since Homer-1b occurs in dendritic spines, the local retention of mGlurs in the ER it induces may modulate the function of these receptors (Roche et al. 1999). Shank thus may play a coordinating role in the signaling mechanisms of AMPArs, NMDArs, and mGlurs (Tu et al. 1999; Sheng and Lee 2001).

MARCK is a scaffolding protein that links the lipid bilayer, protein kinases, and calmodulin and is essential for the control of cell shape (Ramsden 2000).

Spinophilin is a scaffolding protein localized to spine heads. It forms cross-links with the enzyme protein, phosphatase-1, which it directs to its correct location in the spine (Allen et al. 1997).

At the neuromuscular junction, spectrin acts as a scaffolding protein and links via ankryn to the cytoskeletal membrane; it also binds voltage-dependent sodium channels and NCAM into one functional complex (Kordeli 2000).

The beta adrenergic receptor interacts with a PDZ domain-containing protein Na^+/H^+ exchanger regulatory factor (NHERF), or ezrin-radixin-moesin-binding phosphoprotein (EBP50), which mediates recycling of the receptor after endocytosis. Mutation of this motif results in the inhibition of recycling and missorting of the receptor to lysosomes (Garner et al. 2000).

These scaffolding proteins optimize postsynaptic cascades by reducing the distances between the elements of the cascade, preventing these elements from wandering away, and possibly by correctly orienting the interacting molecules themselves (Dimitratos et al. 1999). A postsynaptic cascade is like a conveyor belt, and the scaffolding protein is simply an efficient conveyor belt organizer. Figure 3.5 presents some scaffolding proteins in outline fashion.

Recently Burack and Shaw (2000) have emphasized that scaffolding proteins may have much more dynamic properties than simply acting as inert scaffolds for other molecules. Moreover, Meyer and Shen (2000) distinguish between static scaffolding proteins of the type we have discussed so far and dynamic systems in which the "scaffold" is not a fixed structure, but is a dynamic assembly itself, with ongoing transport of its components to which the enzymes, etc. bind.

Other proteins that link receptors to the cytoskeleton are alpha-actinin-2 and spectrin (Matus 1999). The calcium influx that follows

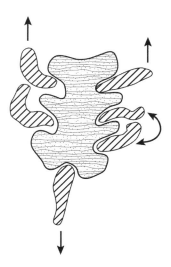

Figure 3.5
Scaffolding protein.

NMDAr activation destabilizes the link between the NMDAr NR1 sub-unit and alpha-actinin-2, resulting in changes in the cytoskeleton that translate into changes in the shape of the dendrites and spines. This calcium inflow also depolymerizes actin and in this way further contributes to changes in spine shape (Matus 1999). NMDAr activation also activates the protease calpain, which cleaves spectrin, leading to disassembly of the cytoskeleton (Bahr 2000). Recently the motor protein myosin Va has been found in PSDs (Walikonis et al. 2000) and may play a role in local transport.

The NMDAr-induced calcium influx is terminated by Ca^{2+}-calmodulin-dependent inactivation that results in a competitive displacement of cytoskeletal-binding proteins from the NR1 subunit (Lu et al. 2000). The NR1 C-terminal can be phosphorylated by PKC and this upregulates this inactivation. On the other hand, PKC phosphorylation can also enhance NMDA-evoked currents, but indirectly [via activation of CAK/Pyk2 and the nonreceptor kinase Src (Lu et al. (2000)]. To complicate matters further, the Ca^{2+} influx also activates the phosphatase, calcineurin, which downregulates the NMDAr activity. This can be counterbalanced by the Ca^{2+} influx activating PKA and/or casein kinase-II (Lu et al. 2000).

With respect to all these complicated interactions, Allison et al. (2000) have pointed out that many proteins bind to cytoskeletal elements in vitro, but they ask "which of these interactions are important for localization of the protein in the synapse?" (Allison et al. 2000, p. 4545). By pharmacological manipulations, they showed that the synaptic clustering of the NMDA-anchoring protein PSD-95 and GKAP (and PSD-95 interactive protein) is independent of the actin and tubulin cytoskeleton. The same is true for the GABA-A receptor and its related clustering protein, gephyrin.

In contrast, the synaptic clustering of Ca^{2+}-calmodulin-dependent protein kinase-II and the actin-binding proteins, debrin and alpha-actinin, are completely dependent upon an intact actin cytoskeleton. Allison et al. (2000) conclude that PSD-95, GKAP, and NMDArs form a stable complex independent of their links to the cytoskeleton. They state that a large number of factors cooperate to determine spine morphology. For example:

- Stabilization of actin filaments by alpha-actinin, spectrin, alpha-adductin, neurabins, and cortactin
- Spine elongation by debrin
- Cleavage of spectrin by calpain
- Inhibition of alpha-adductin function by PKC phosphorylation
- Enhanced synaptic localization of cortactin by glutamate activation

They also note that microtubules are not found in spines, yet their constituent protein, tubulin, occurs in the PSD. They suggest that tubulin here must have some other function. They also speculate that actin filaments at the AMPAr are more concerned with endocytosis and recycling than with anchoring. In general they conclude that actin and tubulin in the spine may be more concerned with activity-dependent changes in spine morphology and signaling than they are with any anchoring function.

It is of interest that the serum in Alzheimer's disease contains anti-brain spectrum antibodies that are not found in normal sera (Fernández-Shaw et al. 1997).

CaMKII may regulate an initial phase of synaptic plasticity that includes translocation target trapping and translocation pruning (for details of the complex chemistry involved, see Shen et al. 2000 and Fukunaga and Miyamoto 2000).

3.2.1 Clustering Molecules
A related group of clustering molecules (e.g., gephyrin, agrin) is involved in synaptic function. MAP-1B binds the GABA receptor to microtubules. Gephyrin is involved in clustering of the glycine receptor (Wheal et al. 1998) and GABA receptors (Simbürger et al. 2000). It is distributed in dendrites in large clusters (associated with parvalbumin) and small clusters (associated with calbindin). Synaptic reorganization following deafferentation results in the redistribution of gephyrin into very small clusters (Simbürger et al. 2000). Other proteins involved in clustering GABArs are GABA receptor-associated protein (GABArAP), collybritin, and dystrophin (Sassoè-Pognetto and Fritschy 2000). Agrin is synthesized in the presynaptic neuron and is excreted and deposited in the synaptic extracellular matrix, where it induces postsynaptic differentiation and neurotransmitter receptor clustering mediated by the scaffolding protein, rapsyn (Koulen 1999; Han et al. 2000). Agrin also induces phosphorylation of the transcription factor CREB in hippocampal neurons via activation of tyrosine kinases in a calcium-dependent manner (Ji et al. 1998).[4]

3.3 Axon Growth-Directing Proteins

The growth of axons during their development is guided by a series of attractant and repellent molecules. Thus the growth of new synapses during adult neuroplasticity may also involve the action of these or similar compounds. Such molecules include netrin, laminin, spectrin, semaphorin, tenascins/ephrins, and scatter factors (Bovolenta and Fernaud-Espinosa 2000; Albright et al. 2000). These growth-directing factors may be either diffusible or substrate bound. Growth cones of retinal axons are attracted by netrin (via its specific receptor *deleted in colon cancer*, DCC, modulated by cytoplasmic calcium levels) and will turn toward a source of it (Yee et al. 1999; Hong et al. 2000). Netrin has been shown to be a specific guidance cue, as well as a survival factor, in the inferior olive during development (Bloch-Gallego et al. 1999). Laminins themselves are potent stimulators of neurite out-growth (Powell and Kleinman 1997), whereas laminin-1 also converts netrin-mediated attraction into repulsion (Höpker et al. 1999).

Semaphorins are a family of growth-repulsant, axon-guidance mole-cules acting on their receptors, neuropilin-1, neuropilin-2, and plexins (Artigiani et al. 1999; Chen et al. 2000; Giger et al. 2000; Raper 2000). Semaphorin H can also act as a stimulator of neurite outgrowth by activation of the MAPK pathway in a calcium-dependent manner (Sa-kai et al. 1999). These chemorepulsant factors such as semaphorin and collapsin can also induce apoptosis (Shirvan et al. 1999). Shirvan et al. suggest that before their death, apoptosis-destined neurons can secrete destructive axon-guidance molecules that can affect the neighboring cells and thus transfer a "death signal" across specific and susceptible neuronal populations.

Wong et al. (2000) suggest that ephrins and semaphorins may act as brakes ("stop signals") to arrest dendritic motility after a viable syn-apse has been formed. Recent evidence (Drescher 2000) suggests that ephrins are not just axon-guiding molecules but are also intimately concerned with the initiation of synapse assembly—in particular eph-rin B and NMDA receptors.

Scatter factors and secreted semaphorins are diffusible ligands, whereas membrane-bound semaphorins signal by cell–cell interaction (Artigiani et al. 1999). Growth-directing factors, such as axonin-1/ TAG-1, can interact with CAMs (such as NCAM) in the differentiation and activation of axonal growth cones in various ways (Stoeckli et al. 1996; Esch et al. 1999). The molecular mechanism here probably uses a form of zipper in which the interacting elements are not individual amino acids but are backfolded polypeptide modules (Freigang et al. 2000). Semaphorin induces dramatic polymerization of actin fila-ments, with very little effect on microtubules (Fritsche et al. 1999). Ming et al. (1999) present evidence that there are common cytoplasmic sig-naling pathways for two separate groups of guidance cues (netrin-1, BDNF, and myelin-associated glycoprotein, MAG) and (collapsin-1-semaphorin-IIID and neurotropin-3), one of which requires coactiva-tion of PLCα and PI-3K pathways. Another recently described group of axon-guidance molecules is the TOAD/ulip/CRMP (TUC) family, which interact with collapsin-1 (Quinn et al. 1999).

3.4 Role of Neurotropins

There are many neurotropic factors involved in synaptic plasticity. The four major neurotropins are nerve growth factor, brain-derived neurotropic factor, neurotropin-3 (NT-3), and NT-4 (Davies 2000). These have their own specific receptors (tyrosine kinase A, B, C, and D, respectively) and another receptor P75^{NTR}, which is common to all. NGF is released constitutively (i.e., continuously) by hippocampal neurons, whereas brain-derived neurotropic factor is released acutely upon activation from dense-core vesicles. NT-3 can function in both ways (Farhadi et al. 2000).

Brain-derived neurotropic factor modulates gene expression, neuronal activity, neurotransmitter release, and synaptic plasticity (Black 1999; Gottschalk et al. 1998). It plays an important role in maintaining the functional activity of dopaminergic mesencephalic, cholinergic septal, cortical and hippocampal, spinal motor, and serotoninergic neurons (Angelucci et al. 2000). BDNF is synthesized and released by pyramidal cells in the cortex. It binds to its TrkB receptor on the postsynaptic neuron, activating various phosphokinases (Levine and Cepeda 1998). These kinases phosphorylate subunits NR1 and NR2 of the NMDAr, which increases the probability of the open state of the ion channel and so boosts performance. Brief depolarizations are sufficient to induce accumulation of BDNF and TrkB mRNAs in the dendrites of hippocampal neurons (Righi et al. 2000).

On the other hand, BDNF has been reported (Rutherford et al. 1998) to inhibit pyramidal cell activity and to augment inhibitory intraneuronal cell activity. During periods of low pyramidal cell activity, BDNF production falls and synaptic strengths are adjusted to promote increased pyramidal cell activity. Conversely, during periods of high pyramidal cell activity, BDNF production increases and synaptic strengths are adjusted to promote interneural inhibitory cell activity. BDNF strengthens neural excitation mainly by augmenting the amplitude of AMPA receptor-mediated miniature excitatory postsynaptic currents (EPSCs), but enhances inhibition by a different mechanism—by increasing the frequency of miniature inhibitory postsynaptic currents (mIPSCs) and increasing the size of GABAergic synaptic terminals (McLean Bolton et al. 2000). BDNF also protects neurons by inhibiting NOS production by reducing NOS induction (Estévez et al. 1998). Thus the system operates as a negative-feed homeostatic mechanism that helps protect against a major problem in the brain—glutamate neurotoxicity.

BDNF activation of its TrkB receptors triggers a cascade involving MAP kinase, which modulates the synthesis of new proteins, including the AMPAr protein, and promotes synapsin phosphorylation and thus glutamate release (Jovanovic et al. 2000). This system plays a role in mediating LTP (Sermasi et al. 2000). BDNF raises AMPAr levels, but not AMPAr mRNA levels. Therefore its effects on this aspect of neuroplasticity are mediated at the translational or post-translational levels (Narisawa-Saito et al. 1999). The growth of dendrites induced by BDNF occurs only in active dendrites. This offers a mechanism by which BDNF selects active neurons for growth (McAllister et al. 1996). In culture, BDNF induces an increase in spine number after 2 weeks. The BDNF receptor molecule TrkB itself induces longer-necked spines, but no increase in spine numbers (Shimada et al. 1998).

Horch et al. (1999) report that BDNF induces structural instability in dendrites and spines, which may help translate activity patterns into specific morphological changes. They suggest that one of BDNF's functions may be to get rid of old connections in order that new ones can form. Lein and Shatz (2000) present data that indicate that BDNF plays a highly local role in strengthening and maintaining active synapses during the formation of ocular dominance columns in primary visual cortex. With the exception of NT-3, all neurotropins affect the outcome of monocular deprivation, but by different mechanisms. NT-4 and NGF counteract monocular deprivation effects without causing detectable alterations in either spontaneous or visually evoked neuronal activity. BDNF is less effective on ocular dominance plasticity, but strongly affects spontaneous and visually evoked activity in cortical neurons (Lodovichi et al. 2000).

Ernst et al. (2000) suggest that during development NO is involved in the elimination of misdirected retinal axons, whereas BDNF in conjunction with NO may stabilize correctly targeted axonal arbors. In the dentate gyrus of the hippocampus, the granule cells synthesize two neurotropins, BDNF and NT-3. The former modulates the synapses formed by the lateral perforant path and the latter modulates the synapses formed by the medial perforant path (Asztely et al. 2000). The perforant path carries the main input to the hippocampus from the entorhinal area.

Recently some proteins have been discovered that are essential for the development of synapses. These include futsch (the *Drosophila* homolog of MAP-1B) (Roos et al. 2000; Hummel et al. 2000) and hiw (a product of the gene, highwire). Its homolog in humans is pam protein (Wan et al. 2000). There are three types of MAPs in neurons: low molecular weight (kinesins, dynamin, MAP-2c, and tau), intermediate (MAP-3 and MAP-4), and high molecular weight (MAP-1A and MAP-1B, MAP-2, dynein). Tau occurs in axons, MAP-2 in dendrites and soma, and MAP-1A and B occur in dendrites, axons, and soma (Hummel et al. 2000). I discuss NGF later.

3.5 Role of Actin

The growth of new spines requires new cytoskeleton as well as new membrane at that site. The shape of a cell or of a spine depends largely on the degree of polymerization of actin, the principal protein of the cytoskeleton. Spines are rich in actin, and direct visualization shows them to change their shape within a few seconds (but not their size in this time frame). The state of actin polymerization also influences the channel properties of the NMDAr (Fischer et al. 1998). The calcium influx induced by NMDAr stimulation also depolymerizes actin and leads to changes in the shape of the spine (Matus 1999). Beta adrenergic stimulation causes rapid neurite outgrowth. This does not require gene expression or protein synthesis, but is due to a rapid and reversible redistribution of filamentous actin (Kwon et al. 1996). The reorganization of the cytoskeleton in response to growth factors is mediated by Rho GTPases that function as molecular switches cycling between an inactive GDP-bound state and an active GTP-bound state (van Aelst and D'Souza-Schorey 1997).

Halpain (2000) has suggested that the actin in spines occurs in two pools. The first pool forms a stable core in the center of the spine shaft and is kept together by barbed and pointed-end capping proteins (figure 3.6). The second pool consists of dynamic actin under the membrane. The latter is linked to membrane proteins by spectrin, to the NMDAr by alpha-actinin, and to the stable pool of actin by debrin. Movements in the dynamic pool are responsible for spine twitching.

3.6 Role of Local Protein Synthesis

Synaptic plasticity is modulated in part by synthesis in the dendrites of new proteins derived from translation of local mRNAs (Huang 1999). This is thought to be mediated in the local organelle called the "spine apparatus" (Lüscher et al. 2000). Local protein synthesis is not required for the initiation of synaptic growth, but it is required for its stabilization and persistence (Casadio et al. 1999). There are three types of targeting of protein synthesis: (1) nuclear and cell wide, (2) nuclear targeted to small domain(s) of synapses, and (3) local protein synthesis. Schuman (1997, 1999) has suggested the following process:

1. After activation of a particular synapse, a signal is sent to the nucleus of the postsynaptic cell to increase transcription and translation of specific proteins.

2. A second signal is generated to mark the synapse that had been activated so that in future it will recognize these newly synthesized mRNAs ("mRNA hijacking").

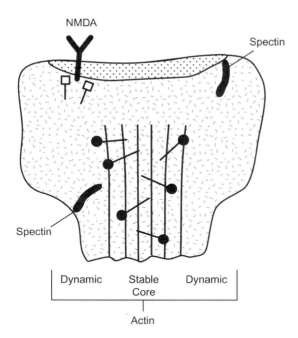

Figure 3.6
Spine actin—stable and dynamic portions.

3. Protein synthesis is stimulated from these mRNAs at the selected synapse.

MRNAs for the AMPAr are, however, confined to the cell soma.
Therefore synaptic targeting for the AMPAr requires long-range protein transport. In contrast, mRNAs for the NMDAr are distributed dif-
fusely throughout the dendrite (O'Brien et al. 1998).

Scheetz et al. (2000) have found that NMDAr activation has two
rapid effects. (1) It induces (within minutes) an increase in the level
of transcription of alpha-Ca^{2+}-calmodulin-dependent kinase II, but a
decrease in the level of total protein synthesis. (2) It induces an increase
in Ca^{2+}-dependent phosphorylation of eukaryotic elongation factor-2
(eEF-2), a process that inhibits peptide chain elongation during protein
synthesis. Scheetz et al. (2000) suggest that the regulation of peptide
elongation by eEF-2 can link NMDAr activation to local increases in
the synthesis of specific protein during activity-dependent synaptic
changes.

They go on to suggest that the calcium influx following NMDAr activation activates eEF-2 kinase and leads to the phosphorylation of

eEF-2. This slows the rate of protein translation and elongation rather than its initiation. This would have the effect of upregulating the translation of abundant but poorly initiated transcripts such as alpha-CaM-KII in dendrites.

Pierce et al. (2000b) point out that dendrites and spines contain protein translocation sites (translocons), ribosomes, and lumenal proteins of the endoplasmic reticulum that are involved in protein assembly and folding. This allows local protein synthesis to be involved in synaptic-specific modification. mGlur activation increases protein synthesis and spine growth. One mechanism for this is via sequential activation of PLC, PKC, translocation of multifunctional p90vsk to polysomes, and increased synthesis of fragile $\times$ mental retardation protein (FMRP), an mRNA-binding protein.

Chapter 4
Miscellaneous Items

4.1 Volume Transmission and Spillover

Nonsynaptic signaling is important in the brain, especially during neurogenesis and in synaptic plasticity. Sesack et al. (1990a) state that dopamine and glutamate can interact by nonsynaptic mechanisms possibly following diffusion from synaptic sites of release. The extensive roles that catecholamines play in volume signaling and that freely diffusible molecules such as NO, hydrogen peroxide, and arachidonic acid play in synaptic signaling are reviewed elsewhere in this book. Other examples:

- Increased neuronal activity in the CA1 area of the hippocampus leads to increased levels of ROS and RNS. The hydrogen peroxide and nitric oxide produced diffuse to neighboring oligodendrocytes and modulate post-translational synthesis of myelin basic protein (Atkins and Sweatt 1999).
- Stimulation of local acetylcholine receptors on dendrites alters the properties of the dendritic membrane along its entire length (Mednikova et al. 1998).
- Activation of the dopamine D1 receptor activates PKA, which results in phosphorylation of the AMPAr Glu R1 subunit at Ser-845, but not Ser-831. Cocaine and methamphetamine also induce phosphorylation of this Ser-845 (Snyder et al. 2000).

Volume release is related to the phenomenon of "spillover." Most discussions of spillover center on the ability of glutamate to diffuse out of its own synapse and to affect neighboring glutamate synapses. This process of cross-talk between neurons is reviewed by Lozovaga et al. (1999). Bergles and Jahr (1997) and Bergles et al. (1999) present evidence for the rapid diffusion of glutamate out of its synaptic cleft in significant amounts. This is normally limited by glia that form a cuff around the synapse. In the cortex and hippocampus, most synapses have gaps in this cuff and many are not well insulated. However, astrocytes grow around active synapses. This may represent an attempt to "plug the leak" or it may represent some quite other function related to the many different ways in which astrocytes and neurons interact (Ventura and Harris 1999). Cuffing also occurs in the neurohypophyseal neurosecretory system.

Stimulation of magnocellular neuroendocrine cells (MNC) leads astrocytes actively to retract their cuffing processes from between MNC and posterior pituitary axons. This leads to the uncovering of postsynaptic sites and to the establishment of additional contacts in the form of novel multiple synapses where one presynaptic axon contacts two or three spines. This is also associated with the formation of dendritic bundles and gap junctions in which tenascin and Nr-CAMS are involved (Laming et al. 2000).

Another example of the active role of astrocytes in synaptic plasticity is found in the supraoptic nucleus (SON), where they change their orientation (vertical versus horizontal), depending on the hydration state. This may modulate communication between SON neurons and the CSF. Similar astrocyte dynamics occur in various hypothalamic neurons related to reproductive functions (Laming et al. 2000). During exposure to a complex environment, astrocytic processes directly apposed to synapses expand their surface area. Laming et al. (2000) suggest that the evidence suggests that astrocytes are not simply packing molecules filling in the gaps between neurons but "appear to be selectively participating in synaptic formation and plasticity" (Laming et al. 2000, p. 326). Motor learning leads to similar changes in astrocytes in the cerebellum. Many neurotransmitters may be involved in astroneuroplasticity (Laming et al. 2000).

Whenever glutamate reuptake fails (as in hypoxia), the excess glutamate spills over onto the presynaptic regions of adjacent Gly synapses and inhibits Gly release. This interference with inhibitory input will exacerbate glutamate neurotoxicity. According to Isaacson (1999), there is strong evidence for spillover of glutamate released from the dendrites of mitral cells in the olfactory bulb. There is also evidence for spillover at the NMDAr, but not the AMPAr (Kullman and Asztely 1998; Rusakov and Kullman 1998). In the cerebellar cortex, spillover has been demonstrated from Golgi cell terminals to granule cell dendrites (Rossi and Hamann 1998). However, glutamate spillover is much reduced at body temperature (Asztely et al. 1997).

Batchelor and Garthwaite (1997) report the discovery of a new neurocomputational device that allows the temporal and spatial integration of signals. In the cerebellum, climbing fibers stimulate AMPArs on the Purkinje cell and parallel fibers stimulate mGlurs on these cells. A single stimulation of a climbing fiber leads to increased intracellular calcium (via opening of the calcium channel controlled by the AMPAr).

This raised intracellular calcium results in increased responses, for a period, to subsequent activation of the mGlur by the parallel fibers. Thus these two inputs to the Purkinje cell are integrated over time by this mechanism.

From the point of view of this book, the possible spillover from dopamine terminals into adjacent glutamate synapses is of more interest than spillover of glutamate into adjacent glutamate synapses. There is no direct evidence that dopamine spills over in this manner. However, there is some indirect evidence in that dopamine neurotoxicity is mediated by the toxic effect of its o-quinone metabolites acting on the NMDArs, not on its own receptors (Michel and Hefti, 1990; Cadet and Kahler 1994; Ben-Shacher et al. 1995; Lieb et al. 1995; Ohmori et al. 1996). This suggests that dopamine, or at least its o-quinone metabolites, must be able to reach NMDArs.

4.2 Other Neurotransmitters

Other neurotransmitters besides glutamate play a role in synaptic plasticity.

4.2.1 Dopamine and Other Catecholamines

The relationships between dopamine release and positive reinforcement and the relevance of the antioxidant properties of dopamine were reviewed earlier. There is evidence that catecholamines act largely by volume transmission; i.e., following release, they diffuse through the neuropil and affect distant receptors (Pickel et al. 1997). Ninety-five percent of varicosities on norepinephrine (as well as 5HT) systems are nonsynaptic boutons-en-passage. Every neuron in the cortex lies within a maximum of 30 μm from a nonsynaptic NE bouton (Dismukes 1997). The dopamine system has a similar arrangement. Dopamine transporter (DAT) uptake sites are distant from dopamine release sites, again underlining the likelihood of diffusion. This is more marked in the prefrontal cortex than in the striatum (Sesack and Pickel 1998). However, a few (3.6%) dopamine afferents in layers IIIb–IV (but not layers I–IIIa) of the prefrontal cortex do target specific parvalbumin-containing GABAergic neurons (Sesack et al. 1998b).

It is interesting that phorbol esters cause a rapid endocytosis of DAT molecules via phospholipase C activation and a clathrin-dependent mechanism. All these DAT molecules are targeted for degradation by the endosomal–lysosomal pathway, which is effected by 2 hours of protein kinase C activation (Daniels and Amara 1999).

Many dopamine D1 receptors at large glutamate synapses do not have any directly attached dopamine axon terminals (Smiley et al.

1994). Dopamine is released from the cell bodies and dendrites of neurons in the ventral tegmental area where there are dendrodendritic synapses (Nirenberg et al. 1997; Rice et al. 1997). The actual terminals of dopamine axons release glutamate, not dopamine (Sulzer et al. 1998). Dopamine neurons release dopamine only from their boutons-en-passage. In the case of 5HT, the situation is somewhat different in that the terminals release both serotonin and glutamate (Sulzer et al. 1998).

It is interesting that the adrenaline system in the brain is different. The cell bodies are located in the C1–C3 groups in the medulla. Adrenaline in C1 is colocalized with glutamate and in the other regions with GABA. Reuptake also occurs outside the synaptic cleft, indicating a degree of volume transmission (Pickel et al. 1997). However, most adrenergic terminals are synaptic and there are few boutons-en-passage. The terminals release adrenaline plus either glutamate (C1) or GABA (C2 and 3).

It used to be thought that the adrenergic system in the brain performed only low-level autonomic functions. Now, however, it has been found that the C1–C3 nuclei have a massive projection to the medial thalamus, nucleus accumbens, and other higher limbic structures, so this system may be involved in much more interesting higher functions (Lew et al. 1977; Herbert and Saper 1992; Otake et al. 1995; Nagatsu et al. 1998; Rico and Cavada 1998), including psychological stress (Otake et al. 1995).

Dopamine plays a role in synaptic maintenance. Dopamine D1 receptor agonists raise the synaptic density of the prefrontal cortex (Sugahara and Shiraishi 1999). In the striatum, dopamine denervation leads to loss of dendritic spines. Dopamine also potentiates the survival of activated synapses and the elimination of nonpotentiated synapses (Arbuthnott et al. 1998). NMDAr responses are augmented by D1 agonists, and non-NMDAr responses are diminished by D2 agonists (Levine and Cepeda 1998). Part of the effect of dopamine on synaptic plasticity may be mediated by its "ordinary" receptor-mediated cascades involving cyclic nucleotides as second messengers. However, redox mechanisms may also be involved (Smythies 1997, 2000).

Dopamine and other catecholamines have been shown to be direct antioxidants and free-radical scavengers (Liu and Mori 1993). Dopamine's antioxidant properties depend on recycling between dopamine and dopamine quinone. This is supported by the report that dopamine prevents cell death in tissue culture by a direct antioxidant mechanism that does not involve dopamine receptors (Iacovitti et al. 1999). Yen and Hsieh (1997) have confirmed this direct chemical property of dopamine involving scavenging ROS.

A second antioxidant mechanism for dopamine is provided by the fact that activation of D2 receptors induces the synthesis of new antioxidant protein, possibly superoxide dismutase (Sawada et al. 1998; Iida et al. 1999), as well as glutathione synthase, gamma-glutamylcysteine synthase, glutathione peroxidase, glutathione reductase, and glutathione S-transferase (Tanaka et al. 2001). Glutathione S-transferase protects against dopamine neurotoxicity, probably by preventing oxidation of dopamine to o-quinones (Baez et al. 1997; Weingarten and Zhou 2001). However, it may also inhibit dopamine-induced apoptosis via action on Jun-terminal kinase (Ishisaki et al. 2001).

A third possible mechanism by which dopamine may exert an antioxidant effect is via dismuting iron–catecholamine complexes. Zhao et al. (1998) and Siraki et al. (2000) have found that dopamine and iron form a redox cycling complex in vitro based on interconversions of ferrous and ferric iron and of dopamine and dopamine quinone. This system acts as a potent dismuter of superoxide, converting five molecules of superoxide into two molecules of water and three molecules of hydrogen peroxide. The question then arises of whether this system occurs in vivo as well as in vitro and if so, where.

One approach is to determine where in the neuron free iron, dopamine, and superoxide could come into physical contact. In this context it may be significant that the transferrin receptor in the external membrane, upon binding a molecule of iron, is also endocytosed together with its cargo and is trafficked to the same endosome to which the dopamine D1 receptor is trafficked. Inside the endosome, free iron is released from the complex. So, if dopamine is also endocytosed together with its D1 receptor, it too would be released inside the endosome and free iron and dopamine could come into physical contact. The source of the superoxide could be the nearby mitochondria, which convert 5% of the oxygen they consume into superoxide. It is therefore microanatomically possible, but still only theoretically, that a dopamine–iron complex may play an important dismuting antioxidant role inside the neuron. Further research is needed to determine if such complexes do indeed occur in vivo.

The postsynaptic cascades for the dopamine receptors and the glutamate receptors are interlinked (figure 4.1). They both modulate phosphorylation of the dopamine and adenosine 3',5'-monophosphate-regulated phosphoprotein (DARPP), in the case of dopamine via cAMP and of glutamate via calcium. DARPP then activates the phosphorylation, via protein phosphatase-1, of a wide range of targets, including receptors (Greengard et al. 1999).[1]

In addition to its antioxidant effect, dopamine is also easily oxidized to neurotoxic o-quinones (as are the other catecholamines) (figure 4.2).

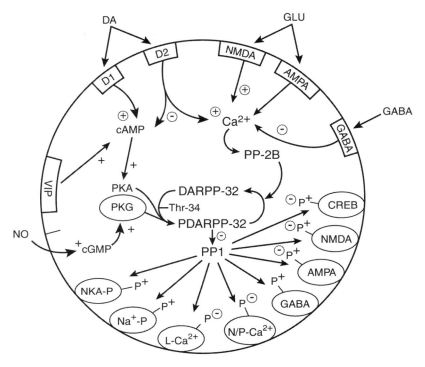

Figure 4.1
DARPP and related molecules. DARPP dopamine and cyclic adenosine 3′,5′-mono-phosphate-regulated phosphoprotein. PP-2B, protein phosphatase-2B (calcium and calmodulin dependent); PKG, cGMP-dependent protein kinase; PKA, cAMP protein kinase A; Thr-34, threonine-34; CREB, cyclic-AMP responsive element-binding protein; DA, dopamine; D1, D2, receptors; NKA-P, neurokinin clathrin adaptor protein; VIP, vasoactive intestinal peptide; NMDA, AMPA, cAMP, cGMP, and GABA as defined in text.

In tissue culture, low doses of catecholamines are neuroprotective via ROS scavenging and high doses are neurotoxic via *o*-quinones (Noh et al. 1990a, Yen and Hsieh 1997). Dopamine *o*-quinones have a variety of effects that lead to neurotoxicity:

> • Catecholamine *o*-quinones are the mediators of catecholamine neurotoxicity (oxidized metabolites of the neurotransmitters dopamine, norepinephrine, and adrenaline). They increase the proton leak across mitochondrial membranes by opening the permeability pore. This leads to uncoupling of ATP synthesis from mitochondrial respiration and cell death (Ben-Shachar et al. 1995; Berman and Hastings 1999).

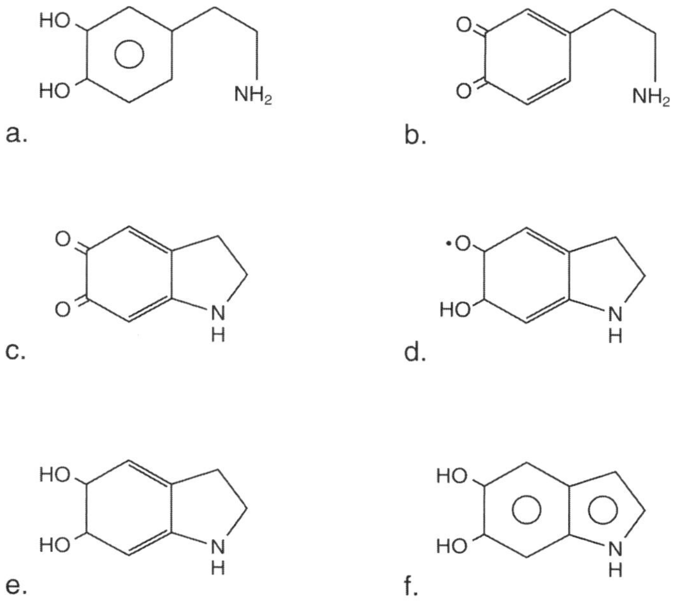

Figure 4.2
Dopamine quinones. (a) dopamine, (b) dopamine *o*-quinone, (c) dopaminochrome (a.k.a. aminochrome), (d) dopamine *o*-semiquinone, (e) dopamine hydroquinone, and (f) 5,6-dihydroxyindole.

- Dopamine *o*-quinones also induce apoptosis via their action on transcription factors such as AP-1 and NF-κB (Luo et al. 1999).
- Dopamine *o*-quinones block the H^+-ATPase proton pump and so inhibit dopamine uptake into synaptic vesicles.
- The free radical *o*-semiquinone directly attacks sulfhydryl groups on proteins.
- Dopamine *o*-quinones have further toxic metabolites, e.g., benzothiazides (Shen and Dryhurst 1996). Dopamine itself also has other toxic tetrahydroisoquinoline metabolites (Maruyama et al. 1995).

By any or all of these mechanisms, catecholamines could, under particular circumstances, be agents of synaptic pruning rather than of the synaptic growth mediated by their antioxidant properties.

4.2.2 Acetylcholine
Intracellular calcium levels, which are of vital importance in synaptic plasticity, are coregulated by NMDArs and α-7 nicotinic acetylcholine receptors. The latter control a calcium channel (Broide and Leslie 1999).

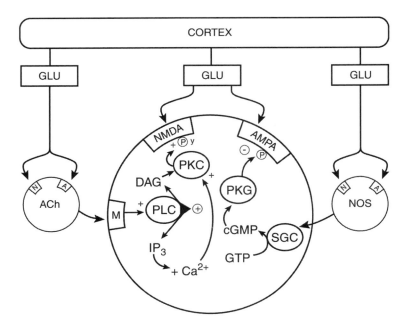

Figure 4.3
Postsynaptic cascade interactions among glutamate (Glu), acetylcholine (Ach), and NOS activity. PLC, phospholipase C; PKC, phosphokinase C; DAG, diacylglycerol; IP_3, inositol-1,4,5-triphosphate; and SGC, soluble guanyl cyclase; N, NMDA; A, AMPA; M, muscarinic acid.

The activation of muscarinic acetylcholine receptors initiates an intraneuronal cascade that directly modulates the cascade produced by stimulation of glutamate receptors (figure 4.3). The muscarinic receptor activates phospholipase C, which converts phosphatidylinositol 4,5-biphosphate to the second messenger, 1,2-diacylglycerol. This activates phosphokinase C, which in turn phosphorylates the NMDAr molecule and upregulates the receptor. In a similar fashion the rise in cGMP levels induced by NO activates phosphokinase G, which inhibits AMPA receptors (Centonze et al. 1999). Further details of the manner of endocytosis of muscarinic receptors are given by Roseberry and Hosey (1999), Edwardson and Szekeres (1999), Vögler et al. (1999) and Bernard et al. (1998).

4.2.3 Serotonin
The serotonin receptor plays an important role in the endocytotic modulation of synaptic function (Hu et al. 1993). 5HT, via its compartmentalized second messenger cAMP, modulates the endocytosis of a set of

neural cell adhesion molecules. It also increases the density of clathrin-coated pits and vesicles and increases the expression of the light chains of clathrin. Hu et al. (1993) conclude that one function of 5HT is to initiate a coordinated program of endocytosis. Serotonin also modulates the release by astroglia of S-100β neurotropic factor, which activates the MAPK-2 pathway to stimulate microtubule production and to cross-link microtubules to cytoskeletal elements, thus stabilizing dendrites (Whitaker-Azmitig et al. 1995).

4.3 Arachidonic Acid Signaling

Arachidonic acid is another freely diffusible signaling molecule (like NO and hydrogen peroxide). It is produced in the post-NMDAr cascade by the enzyme, phospholipase A2, from membrane lipids. It is an important volume transmitter. For example, when a growing mossy fiber approaches a granule cell, the glutamate it releases activates NMDArs on the granule cell. The AA produced by the post-NMDAr cascade diffuses back to the growing tip of the mossy fiber and blocks further growth (Baird et al. 1996) so that the mossy fiber is able to make the synapse with its target (if the appropriate recognition molecules are correctly matched).

Stimulation of the acetylcholine muscarinic receptor also leads to AA release. This exerts feedback inhibition of the acetylcholine receptor and also stimulates guanyl cyclase to produce the second messenger cGMP (Kjome et al. 1998). AA is also produced by the action of NCAM on PLAγ (Doherty et al. 1995). AA also activates potassium channels in rat visual cortex (Horimoto et al. 1997).

Neuronal growth cones are very rich in PLA2 and AA. AA and lipoxygenase metabolites regulate actin-based neurite remodeling (extension and retraction). The enzyme that produces AA, cyclooxygenase 2 (COX-2), is present in large amounts in spines and dendrites, especially in layer II and III pyramidal cells in the dentate gyrus and lateral amygdala (Kaufman et al. 1996). A detailed account of the anatomical distribution of this enzyme in rat brain is given by Breder et al. (1992). COX-2 (one part of PGH synthase) mRNA induction by NMDAr activation in the hippocampus plays a major role in synaptic plasticity (Yamagata et al. 1993). This may depend as much on the ROS produced by this enzyme as on its role in the prostaglandin synthetic pathway. AA as a retrograde messenger in vitro is relatively ineffective if it is applied by itself, but it potentiates the effect of other signaling molecules (Smalheiser et al. 1996). AA also inhibits the GluT acting from the water phase, not from the lipid membrane (Trotti et al. 1995).

4.4 Hormonal Modulation of Synaptic Plasticity

Several hormones have an effect on synaptic plasticity. During the estrous cycle, the number of synapses in the brain fluctuates. In particular there is a very rapid elimination (<24 hours) in the period between proestrus and estrus (Desmond and Levy 1998). Estradiol increases spine density in the hippocampus by reducing local GABAergic neurotransmission. GABA blockers, such as picrotoxin and bicuculline, also increase spine numbers in the hippocampus (Murphy et al. 1998a).

Estradiol also lowers BDNF levels in brain to 40% of control value in 24 hours. This lowers inhibitory tone and increases excitatory tone in pyramidal neurons. A twofold increase in spine density results. This effect is mediated by increased phosphorylation of cAMP response element-binding protein (Murphy et al. 1998b). Estradiol increases the density of spines and synapses in hippocampal CA1 cells. It also increases the NMDAr binding of glutamate and the sensitivity of CA1 pyramidal cells to NMDA-mediated synaptic input (Woolley et al. 1997). In the ventromedial nucleus of the hypothalamus, estrogen increases spine density by 48% in the ventrolateral portion, but not in the dorsal part of this nucleus. Moreover it does so only on short primary dendrites, not on long primary dendrites. The mechanism is not mediated by estrogen receptors and so may be transsynaptic (Calizo et al. 2000).

During lactation, the dendrites that arise near the soma of oxytocin neurons shrink by some 40% in their total length. In vasopressin neurons, there is a 50% increase in total dendritic length and an increase in close-to-soma dendritic branching (Stern 1998).

4.5 Psychological Stress

Psychological stress also affects synaptic plasticity. In rat frontal cortex, repeated mild stress leads to sprouting of norepinephrine axons. Severe stress has the opposite effect and leads to a reduction of NE axons. Psychological stress modulates stress-activated protein kinase activity (Goodman et al. 1998). A forced walking stress carried out for 2 weeks induced a reduced density of noradrenergic axons in rat frontal cortex (Kitayama et al. 1997). Exposure to acute psychological stress in tree shrews caused a rapid reduction in new neurogenesis in the dentate gyrus. In contrast, NMDAr blockade had the opposite effect (Gould et al. 1997).

Adamec et al. (1999) have produced pharmacological evidence that NMDArs are involved in the initiation but not the maintenance of the neural changes that mediate increases in anxiety following severe stress. Stress in a rat model initiated increased PKC activity in the amygdala that was dependent on NMDAr activation. This was mediated by translocation of the PKC molecules from the interior of the cell to the external membrane (Shors et al. 1997).

4.6 Energy

A growing synapse needs not only increased material but also an increased energy supply. Some of the details of this link have been worked out by Magistretti and Pellerin (1999). The connection takes place in astrocytes. The conversion of glutamate to glutamine in the astrocyte requires the conversion of ATP to ADP. ADP is recycled back to ATP by the aerobic metabolism of glucose to lactate. The lactate is then transported to the axon terminal, where it enters the tricarboxylic acid cycle that yields more ATP molecules to supply local energy needs. In this way the energy supply is linked to the level of glutamate release. However, it has also recently been shown that the glutamatergic terminal can also synthesize glutamate from pyruvate (Hassel and Brathe 2000). It was previously supposed that only astrocytes could do this.

4.7 The Role of Astrocytes

Astrocytes do not act just as a passive support system for neurons, as it was long thought they did. They also play an active role in signal transduction. Activation of Glu receptors on astrocytes leads to increased calcium levels inside the astrocyte. Some of the calcium can pass into neighboring astrocytes in the form of calcium waves via the gap junctions that connect astrocytes. This starts the chain of raised AA level, activation of PGH synthase, and raised prostaglandin E_2 (PGE_2) levels.

Prostaglandins are freely diffusible molecules and can have paracrine effects on neighboring astrocytes and neurons. The PGE_2 can also trigger Ca^{2+}-mediated glutamate release from astrocytes, probably by an exocytic mechanism (Vesce et al. 1999). This glutamate released from astrocytes can affect neighboring neurons in various ways, e.g., it can promote slow inward currents via ionotropic glutamate receptors, decrease the amplitude of postsynaptic currents (via mGlurs), and increase miniature postsynaptic currents via extrasynaptic NMDArs (Vesce et al. 1999). The nitric oxide-G kinase signaling pathway has a basic role in calcium waves (Willmott et al. 2000). The mGlurs on the astrocyte membrane activate related adhesion focal tyrosine kinase

(RAFTK) and MAPKs (ERK-1 and 2), leading to modulation of DNA synthesis (Schinkman et al. 2000).

Astrocytes require specific signals from neurons to develop their neuroprotective role. One such example is vasoactive intestinal peptide (VIP). This is released by many neurons following depolarization. It upregulates glutamate reuptake by the astrocyte and expression of BDNF. It also induces the release of the neuroprotective cytokine IL-6 from the astrocyte (Brown 2000).

Astrocytes also play a role in axon guidance. This is effected by changing the patterns of specific extracellular matrix molecules on their surface, such as tenascins, chondroitin, and proteoglycans (Powell et al. 1997).

Chapter 5

Pharmacological Implications and Clinical Applications

5.1 Pharmacological Implications

The endocytosis-recycling system for receptors clearly plays a key role in receptor function and in synaptic plasticity and so offers a new venue for the site of action of known, as well as yet unknown, drugs. For example, the neuroleptic, chlorpromazine, acts on endosomes by inducing a redistribution of a clathrin-coated pit component, AP-2 (Subtil et al. 1994). It also reduces transferrin, but not IL-2 endocytosis; it powerfully inhibits actin polymerization and alters cell shape and motility in lymphocytes (Milzani and Dalledonne 1999).

The neuroleptic, haloperidol, activates the PKA-induced cAMP pathway leading to phosphorylation of the NR1 subunit of the NMDAr at Ser-897. This recruits NMDArs to the site and potentiates NMDAr function. Haloperidol also very rapidly induces the c-fos and pro-enkephalin genes. NMDAr activity is essential for the regulation of gene expression by clozepine and haloperidol, and this may play a role in the antipsychotic effect of these drugs (Leveque et al. 2000b).

Lindén et al. (2000) have measured the effect of antipsychotic and psychotomimetic drugs on the expression of mRNA for, and actual levels of, the neurotropins BDNF and NT-3 in the entorhinal cortex. They found that clozepine had little effect, but the ketamine analog MK-801 induced raised levels of BDNF mRNA (but not NT-3, or their receptors TrkB and TrkC) in the entorhinal cortex. Levels of BDNF were raised in the entorhinal cortex (by 126%, plus or minus 7%) and fell by 71%, plus or minus 2%) in the hippocampus. It is interesting that this effect was attenuated by pretreatment (1 hour) with clozepine and haloperidol, but not by imipramine. Lindén et al. (2000) suggested that these effects might have some relevance for the clinical effects of these drugs. A somewhat different result was obtained by Angelucci et al. (2000), who gave haloperidol by mouth to rats for 29 days and found it to decrease BDNF levels in the frontal and occipital cortices and hippocampus and to raise TrkB levels in selected brain regions.

The action of many CNS-active drugs is currently attributed entirely to their effects on receptors. Additional sites of action may be (1) the mechanisms described in this book by which these receptors are endocytosed, processed, and recycled and (2) the postsynaptic cascades inaugurated by receptor activation. Some preliminary work has been reported in this area. Sojakka et al. (1999) found that isoproterenol inhibits transport from early to late endosomes and causes fragmentation of the late endosome. Henkel et al. (1999) report that substrates for and inhibitors of the intracellular protease Kex2 can be delivered by endocytosis. They suggest that this method might be of therapeutic use in the virus infections that depend on these proteases. Antidepressant drugs upregulate the activity levels of cAMP response-mediated gene transcription and the phosphorylation of CREB protein (Thome et al. 2000).

Whistler et al. (1999) have found that addictive opiates differ from equally effective opiates with low addiction potential in that the latter are endocytosed after they bind to the mu receptor, whereas the former are not. These authors suggest that this finding indicates "a fundamental revision of our understanding of the role of receptor endocytosis in the biology of opiate drug action and addiction" (Whistler et al., 1999, p. 737). Furthermore, morphine fails to cause arrestin translocation to the membrane (Roth and Williams 1999). Roth and Williams introduce a term to express this: relative activity versus ability to induce endocytosis (RAVE). Drugs with low addiction potential such as methadone and etorphine have low RAVE values, whereas drugs with high addiction potential such as morphine have high RAVE values.

5.2 Clinical Applications

Disturbances in the mechanisms underlying synaptic plasticity may be involved in the pathogenesis of several diseases of the central nervous system.

5.2.1 Schizophrenia

In schizophrenia, until very recently research has tended to focus on receptors: at first dopamine receptors and more recently glutamate receptors. However, the discovery that the dendritic spines in the cortex and striatum are reduced by some 50% in the disease (Glantz and Lewis 1995, 2000; Goldman-Rakic and Selemon 1997; Garey et al. 1998) suggests that some abnormality in the system that maintains dendritic spines may be relevant.

Debate at first centered on whether this reduction in spine numbers meant that the spines were not grown in the first place or that they were subject to excessive removal later. However, now that it is known that spines are ephemeral structures, the more likely explanation is that the reduction in numbers is due to a shift in the ongoing dynamic balance between spine formation and spine pruning.

I have suggested elsewhere that one factor controlling synaptic plasticity is the redox balance at the glutamate synapse between neuro-destructive pro-oxidants (such as hydrogen peroxide, superoxide, and the hydroxyl radical) and neuroprotective antioxidants such as ascorbate, glutathione, dopamine, and perhaps carnosine (Smythies 1997, 2000). In this case it may be significant that there are many studies that show that antioxidant defenses are impaired in schizophrenia (Yao et al. 1998a,b,c and see Smythies 1999 for a review). Edgar et al. (2000), using brain proteosome analysis, showed gene 6q-linked alteration of the concentration of Mn-superoxide dismutase (as well as others) in the schizophrenic hippocampus. These authors conclude that these results indicate that antioxidant defense may be altered in the schizophrenic hippocampus.

Recently Do et al. (2000) have reported, using proton magnetic resonance spectroscopy, that in schizophrenia glutathione levels are reduced by 52% in the prefrontal cortex in the living brain as well as by 27% in the spinal fluid. Many studies have established that methionine induces a worsening of the psychosis in many chronic schizophrenics. Methionine inhibits the uptake of cysteine into cells and therefore blocks the synthesis of the main antioxidant compound in cells—glutathione.

A second potential group of agents involved in spine pruning consists of the catecholamine o-quinones. These compounds certainly occur in the brain in particular loci because they are necessary metabolic precursors of the black pigment, neuromelanin. This is found in the cell bodies of the dopamine-containing cells of the substantia nigra and ventral tegmental area, in the norepinephrine-containing cells of the locus coeruleus, and in the adrenaline-containing cells of the C2 group in the medulla. The level of one of their metabolites, 5-cysteinyl dopamine, is raised in the brain in schizophrenia (Carlsson et al. 1994). One of them, adrenochrome, has been reported by four groups to be a psychotomimetic agent in normal volunteers (Hoffer et al. 1954; Schwartz et al. 1956; Grof 1963; Taubman and Jantz 1957).

The enzyme catecholamine-O-methyl transferase (COMT), which mediates all transmethylation reactions in the body, has an important role in detoxifying toxic catecholamine o-quinones. In the metabolism of catecholamines by this route, the first product is the catecholamine o-quinone (see figure 4.2). This then cyclyses to form the aminochrome, which is in equilibrium with the hydroquinone and the highly toxic free radical, semiquinone. COMT O-methylates the hydroquinone to harmless products that are excreted, thus lowering the production of the semiquinone.

In schizophrenia we have shown that two enzymes concerned in transmethylation reactions, methionine-adenine transferase (MAT) and serine hydroxy methyl transferase (SHMT), are underactive (Smythies et al. 1997). Recently Harada et al. (2001) have carried out a study using DNA from 87 schizophrenic patients and 176 normal controls. They found a highly significant genetic defect in the DNA that codes for the enzyme, glutathione S-transferase-1. This enzyme detoxifies catecholamine o-quinones, including adrenochrome, by adding a glutathione moiety at the 5 position (Baez et al. 1997). Harada et al. conclude that a glutathione S-transferase-1 gene deletion may constitute a risk factor for schizophrenia. Thus in schizophrenia there is a combination of defective antioxidant defenses and excessive production of, and failure to detoxify, catecholamine o-quinones, together with defects in the transmethylation mechanism for further reducing their toxicity. All of this would tend to favor spine deletion over spine growth.

There have also been rather inconsistent preliminary reports on some abnormalities in cell adhesion molecules and related compounds in schizophrenia. The most consistent of these is the finding of reduced levels of synaptophysin in the cortex, particularly the prefrontal cortex (Blennow et al. 2000; Davidsson et al. 1999; Glantz and Lewis 1997; Goldman-Rakic and Selemon 1997; Karson et al. 1997; Perrone-Bizzozero et al. 1996). However, Landén et al. (1999) reported raised levels in the left thalamus, and Young et al. (1998) reported raised levels in the granule cell layer of the dentate gyrus (and normal levels elsewhere). In a later paper, Young et al. (2000) found that the increase in synaptophysin levels and the decrease in SNAP-25 levels in the dentate gyrus were concentrated in the layer under the molecular border where the earliest-born granule cells are located. Levels of mRNA for synaptophysin were raised twofold (Tcherepanov and Sokolov 1997).

Levels of the trophic factor, growth-associated protein-43 have been reported as significantly raised in the hippocampus and cingulate gyrus (Blennow et al. 2000; Perrone-Bizzozero et al. 1996). Blennow et al. suggested that reduced synaptic density might elicit reactive increased synaptogenesis. The latter authors point out that GAP-43 is involved in the initial establishment and reorganization of synaptic connections. In neurons, GAP-43 induces neurite growth (Laux et al. 2000). In early life many neurons contain it, but it disappears later in most of them except in the hippocampus and association areas. Therefore the raised levels in these regions in schizophrenia may indicate a perturbed organization of synaptic connections in distinct cortical association areas in the disease. mRNAs for GAP-43 were found to be reduced in the medial temporal lobe and anterior cingulate, but were normal in the prefrontal cortex (Eastwood and Harrison 1998). Rab3a levels were found to be reduced by 50%, with very little overlap with normals in the left thalamus, but not the right (Blennow et al. 1996).

Cotter et al. (1998) reported that levels of beta- and alpha-catenins (components of the Wnt signaling system) are reduced in the hippocampus in schizophrenia. Honer et al. (1997) found that levels of syntaxin and NCAM are raised in the cingulate cortex in schizophrenia. They say that this is consistent with an increased glutamatergic input to the cingulate cortex. NCAM and L1 antigen levels have also been reported to be raised in the CSF (Poltorak et al. 1995, 1997). According to Vawter et al. (1998a), only one type of NCAM is raised in the prefrontal cortex and hippocampus, which is the 105–115-kDa type. Other NCAMs and L1 protein were normal. However, these same authors (Vawter et al. 1998b) expressed the opinion that this result was due to neuroleptics. More recent work by this group (Vawter et al. 2000) reports an increase in the variable alternative spliced exon (VASE), a 10-amino acid sequence inserted into the fourth NCAM domain in the CSF in patients with schizophrenia but not with affective disorders. Levels of mRNAs for synapsin-1A and -1B are raised, according to Tcherepanov and Soklov (1997). SNAP-25 levels are reduced in the hippocampus (Young et al. 1998).

An added complexity was provided by Thompson et al. (1998), who reported that levels of SNAP-25 were reduced in the inferior temporal cortex, and area 10 of the prefrontal cortex, but were increased in area 9 of the prefrontal cortex and were normal in area 17. DeLisi (1997) has suggested that schizophrenia may be due to defective genes related to neurotropins.

Very recently it has been reported that transcripts encoding proteins involved in presynaptic function are reduced in schizophrenic brain. Studies on 250 other gene groups showed that they were normal (Mirnics et al. 2000). The most consistently abnormal were N-ethylamide-sensitive factor and synapsin-II. Novak et al. (2000) reported that mRNA levels of calcium-calmodulin-dependent protein kinase B are reduced in the frontal cortex in schizophrenia. The abnormal microanatomy found in the brain in schizophrenia is not limited to spines. Kalus et al. (2000) studied 3-D Golgi-stained cortical pyramidal neurons and found that the apical dendrites are normal, but the basal dendrites are markedly reduced in total length and in the number of distal segments.

Neuroleptics such as haloperidol given over a long time course result in lower spinophilin levels and increased phosphorylation of MAP-2 in certain cortical areas. This may be related to the causation of tardive dyskinesia (Lidow et al. 2001).

All reports in the field of the pathophysiology of schizophrenia suffer, alas, from a lack of consistency. This is mainly because schizophrenia is a syndrome, not a unitary disease, with many uncontrolled variables and data overlaps in even the best studies. Moreover, the interpretation of the research findings is often difficult. If the level of a certain molecule that promotes synaptic development, for example, a certain CAM, is found to be raised when the reduced connectivity theory of schizophrenia might lead one to expect that the level should have been reduced, the concept of "compensatory" overproduction is invoked. The explanation here is that the basic lowering of synaptic connectivity leads to a local overproduction of certain of the elements involved, just as dopaminergic denervation leads to an increased supply of dopamine receptors in the affected region.

5.2.2 Possible New Roles for Copper and Iron in the Brain

Currently our ideas of the roles of copper and iron in the brain are limited to the role they play in the action of key enzymes. However, they may have other roles. The first clue that suggested other possible roles came from the accidental discovery by Blake et al. (1985) that a mixture of the iron chelator, desferrioxamine (100 mg), and the neuroleptic, prochlorperazine (25 mg), induces a profound and prolonged (2–3 days) coma in normal humans (with complete recovery). The same dose of desferrioxamine and 12.5 mg prochlorperazine in rats produces a coma lasting some 8 hours. Neither drug by itself produced any such effect. In humans, a marked rise in CSF chelatable copper and total iron levels and a fall in chelatable iron accompany the coma.

The hypothesis developed by Blake et al. to explain this result was that desferrioxamine is a hydrophilic chelator of heavy metals, and prochlorperazine, they claimed, is a lipophilic chelator of heavy metals. Thus they suggested that the two together acted synergistically in mediating a flux of copper and iron ions across the cell membrane. This flux they suggested would "disturb" receptors for norepinephrine and serotonin in the membrane and thus, since these neurotransmitters are related to sleep, somehow this flux induces the coma. They pictured the desferrioxamine chelating iron (and copper) inside the neuron and taking it to the cell membrane, where the prochlorperazine would take up the heavy metal ion, transfer it across the membrane, and deliver it to another molecule of desferrioxamine in the cytoplasm on the external side of the membrane. They conducted some experiments using

an artificial chloroform "membrane" separating two aqueous depart-
ments and showed that in that preparation the desferrioxamine-
prochlorperazine mixture did indeed accelerate the transfer of iron ions
from one aqueous compartment to the other across the chloroform
layer.

However, this hypothesis seems highly improbable for the following
reasons:

- There is no evidence whatever that prochlorperazine can chelate
  heavy metal ions. Its side chain looks vaguely like a molecule of
  spermine, which can chelate heavy metal ions, but the status of
  the nitrogens in each is quite different. In spermine it is —NH—
  whereas in prochlorperazine it is —N=. It is highly unlikely that
  the latter is involved in the chelation of heavy metal ions.
- Desferrioxamine is certainly hydrophilic, but it crosses cell
  membranes easily by means of an endocytotic mechanism (Breuer
  1995; Ollinger 1995). So there is no need to invoke another lipophilic carrier.
- A chloroform layer is really quite unlike a cell membrane.
- It is highly implausible to suggest that the movement of prochlorperazine carrying its alleged load of iron across the neuronal
  membrane would selectively "disturb" receptors for NE and 5HT
  and produce coma.
- If the hypothesis of Blake et al. were true, then one would expect
  that iron-deficient rats would have shorter coma times because they would have fewer iron ions to "disturb" the membrane. However, the experiments of Blake et al. showed that the
  coma in iron-deficient rats lasts ~24 hours, not ~8 hours as in
  normal rats.

I therefore suggest that this hypothesis should be abandoned. In its
place we suggest another working hypothesis: that the coma is due to
the severe fall in the intraneuronal levels of iron and copper. This follows the transfer by desferrioxamine of these ions out of the neuron
into the CSF. There may also be some role for prochlorperazine that
does not involve it acting as a chelator of heavy metals itself.

The next clue came from the observation by Peter O'Brien's group
at the University of Toronto that I mentioned earlier (Zhao et al. 1998;
Siraki et al. 2000). They found that catecholamines and iron react in
vitro to form a complex cycle involving the interconversion of ferric
and ferrous ions and catecholamines with their quinones. This system
acts as an effective dismuter of superoxide, converting five molecules
of superoxide per cycle into two molecules of oxygen and three of
hydrogen peroxide. Currently there is no evidence that this system

operates in vivo as well as in vitro, but if it does, it may operate inside the early endosome for reasons mentioned earlier (Smythies 1999, 2000).

There is some evidence to suggest that copper may also be endocytosed by the copper-transporting P-type ATPase menkis protein (MNK) via a clathrin-coated pathway colocalizing with transferrin (Petris and Mercer, 1999) and/or possibly by prion protein C (PrPC) (Pauly and Harris, 1998) and/or the Ctr family of membrane proteins (Lippard, 1999). Copper, either by itself or in a complex with catechols, can also dismute superoxide (P. O'Brien, personal communication).

Thus we can construct a hypothesis to explain the desferrioxamine-prochlorperazine coma. The desferrioxamine removes the iron (and copper) from the neuron. Prochlorperazine, being a blocker of dopamine receptors, cuts off the supply of endocytosed dopamine. This results in the collapse of the O'Brien cycle and leads to superoxide poisoning. This hypothesis can easily be tested since it predicts that other neuroleptics active at the dopamine D1 receptor, such as chlorpromazine, which could never chelate heavy metal ions, would also be active in producing the coma when combined with desferrioxamine.

If the O'Brien cycle does operate in vivo, then it clearly would play a key role in normal cells in combating superoxide toxicity. Mitochondria convert 5% of the oxygen they consume into superoxide. This is somewhat counterintuitive since iron and copper (outside their rigidly segregated protein envelopes) are usually regarded as highly toxic pro-oxidants. The role of iron in particular in promoting the Fenton reaction, which converts hydrogen peroxide into the highly toxic hydroxyl radical, is well recognized. Clearly it is important to find out if the O'Brien reaction does occur in vivo, either by direct biochemical methods, or by indirect ones such as our proposed use of the desferrioxamine-prochlorperazine-induced coma as a research tool.

5.2.3 Alzheimer's Disease

It is widely held that Alzheimer's disease is an inflammatory disease of the brain associated in some way with the excessive production of the powerful oxidant protein, beta-amyloid. This causes oxidative damage to proteins, lipids, and nucleic acids. The endosome system seems to be involved, because the volume of the endosomes in affected cells is twice to three times normal (Cataldo et al. 1997; Nixon et al. 2000). The first intracellular manifestation of the Alzheimer process is this increase in endosome size (Nixon et al. 2000). Nixon et al. also point out that many of the genetic indicators of the risk of developing Alzheimer's disease encode for proteins that depend on endocytosis for their function. This enlargement is not due to the need to process the excessive amounts of beta-amyloid (Nixon et al. 2000). Rather, it may represent an attempt to repair the oxidized membrane proteins cycled from the membrane to the endosome for this purpose.

A role of some disorder in synaptic plasticity is also suggested by the fact that the disease commences in the medial temporal lobe, which has the highest rate of synaptic plasticity in the brain, and then spreads to other association areas, which show intermediate rates of synaptic plasticity. It finally affects the primary sensory and motor cortices, which have low rates of synaptic plasticity. Lysosome activity is also increased in the disease (Nixon et al. 2000).

The next clue came from the findings that levels of copper and iron are significantly raised in the brain in Alzheimer's disease and that the neurotoxicity of beta-amyloid is mediated by the cell-free generation of hydrogen peroxide by the peptide when it binds catalytic amounts of copper (Cu^{2+}) (Cuajungco et al. 2000). These reactions are counteracted by zinc. Furthermore, treatment of beta-amyloid in vitro with the heavy metal chelator, desferrioxamine, leads to loss of neurotoxicity. If the iron is replaced in the medium, neurotoxicity returns (Rottkamp et al. 2001).

Therefore it would seem that an investigation of the therapeutic effects of raising zinc levels or lowering copper and iron levels would be indicated in the treatment of Alzheimer's disease. Since raising zinc levels in the brain is rendered difficult by the rigid control of uptake, a trial of methods to lower iron and copper levels is indicated. It seems reasonable to suggest that research should be directed toward the latter possibility. One obvious candidate to consider would be the desferrioxamine-prochlorperazine combination, but at lower doses than those used by Blake et al. (1985) that induced a prolonged coma. It may well be possible to find a dosage that achieves the significant lowering of intraneuronal iron and copper possibly involved in the reaction reported by Blake et al. (1985) without inducing coma. No direct evidence of lowering of intraneuronal iron and copper levels was provided by their human clinical data, but one can deduce the possibility of this from the raised CSF levels.

Preliminary animal work would show whether the desferrioxamine-prochlorperazine combination in fact lowers intraneuronal iron and copper levels, suggest the optimal dosage level in humans, and provide further details of the biochemical mechanism responsible for the coma. Such animal work would also provide data on the feasibility of long-term treatment with the combination in terms of possible toxicity reactions.

McLachlan et al. (1991) have reported that desferrioxamine by itself has a therapeutic effect in Alzheimer's disease. They conducted a

2-year controlled trial of administering 125 mg of desferrioxamine twice a day for 24 months and found a significant reduction in the decline in living skills in the patients receiving desferrioxamine compared with those receiving a placebo. There were only minor side effects. It remains to be determined if adding prochlorperazine to the desferrioxamine improves this therapeutic effect without producing undesirable side effects.

ADAM-1 and ADAM-2 are membrane-bound metalloproteases that act as receptors for integrin and function as modulators of the mitotic clock in cell-cycle regulation. Cell-cycle alterations are found in neoplasia and have been reported in Alzheimer's disease. Gerst et al. (2000) report that levels of these proteins are raised in the brain in Alzheimer's disease. This would lead to a loss of matrix organization and resetting of the mitotic clock.

In a recent paper, De Ferrari and Inestrosa (2000) have suggested that Alzheimer's disease may be linked with a sustained loss of function in the Wnt signaling pathway. They point out that beta-amyloid activates GSK-3β, which phosphorylates the skeletal protein Tau, hyperphosphorylation of which is characteristic of the disease. The Wnt pathway inhibits GSK-3β activity.

Kokubo et al. (2000) suggest that DIGs may be involved in Alzheimer's disease. DIGs are detergent-insoluble glycolipid- (and cholesterol-) enriched membrane rafts that contain a variety of signaling molecules and are involved in signal transduction and membrane trafficking. DIGs contain beta-amyloid (as well as amyloid-precursor protein and presenilins) and may be a main site of synthesis of beta-amyloid or at least of its transport from the endoplasmic reticulum to the Golgi apparatus. Kokubo et al. also involve flotillin-1 and -2, which are coeval-associated integral membrane proteins abundant in the brain that may substitute in neurons for caveolin, which they say is not found in neurons. Brain levels of flotillin-1 and -2 are raised in Alzheimer's disease (Kokubo et al. 2000).

However, Lah and Levey (2000) suggest that presenilins are located, not in the ER and Golgi apparatus as was previously thought, but in early endosomes, where they are an important site for beta-amyloid production. Presenilins mediate the gamma-secretase cleavage of APP and are also involved in the proteolytic cleavage of the Notch receptor,[1] TrkB and other proteins (van Gassen et al. 2000). Presenilins also modulate capacitative calcium entry (CCE) into neurons, which is a key component of the phospholipase C-inositol triphosphate signaling pathway. The mechanism is that PLC action liberates IP-3, which binds to specific receptors or channels in the endoplasmic reticulum. This results in the discharge of stored calcium, which activates plasma membrane calcium channels, leading to CCE modulation (Yoo et al. 2000). Yoo et al. conclude that the altered storage of calcium in the endoplasmic reticulum could have a profound effect on many proteins in relation to the pathogenesis of Alzheimer's disease.

Bothwell and Giniger (2000) suggest that the pathobiochemistry of Alzheimer's disease may involve some mechanisms that are prominent during neuronal development and that guide newly formed neurons from the germinal zone to the Cajal–Retzius layer, where differentiation takes place. These agents include reelins, cadherins, the apoenzyme E (ApoE) receptor 2 and the very low-density lipoprotein (VLDL) receptor, the protein "disabled" (commonly known as DAB) and the Ser/Thr protein kinase, cyclin-dependent kinase (CDK5) and its proteolytic fragments p35 and p25. Levels of p25 are raised in degenerating parts of the Alzheimer brain. p25 activates CDK5 activity, which leads to hyperphosphorylation of Tau (see Bothwell and Giniger 2000 for details).

Alzheimer's disease is characterized by extensive degeneration in the cholinergic cells in the basal forebrain (nucleus of Meynart). This may be related to the fact that beta-amyloid binds selectively and with picomolar affinity to alpha-7 nicotinic acetylcholine receptors (Wang et al. 2000). Wang et al. point out that beta-amyloid is selectively toxic to cells that bear this receptor. They also state that beta-amyloid binds to scavenger proteins such as RAGE (receptor for advanced glycation end products), alpha-2-macroglobular protein, and ApoE.

It is widely held that insoluble beta-amyloid deposits represent a toxic agent in Alzheimer's disease. However, there is also evidence that they may normally be protective, acting as a superoxide antioxidant (Bush et al. 2000). Moreover, neurons that contain neurofibrillary tangles have very low levels of 8-hydroxyguanosine (which is evidence of antioxidant damage) whereas they show obvious evidence of previous oxidative damage (advanced glycation end products and lipid peroxidation) (Smith et al. 2000). Smith et al. suggest that this shows that both beta-amyloid and neurofibrillary tangles may be cellular antioxidant compensations for the raised levels of oxidative stress.

Recently Yoo et al. (2001) have presented evidence that Alzheimer's disease is accompanied by deranged expression of chaperone proteins in the brain. These include several heatshock proteins. These authors link this finding to the depressed rate of glucose metabolism, glycogen accumulation, fall in the level of key amino acids, NH_3 toxicity, and Ca^{2+} disregulation found in the brain in Alzheimer's disease.

Chapter 6

Conclusions

It will be evident from this review that the biochemical basis of synaptic plasticity is enormously complex. This is not only because of the very large number of systems involved but also because these systems interact with each other in so many different ways. Moreover, a molecule may have one effect on a system under one set of circumstances, but the opposite effect under a different set of circumstances.

What are the key elements in determining whether a certain synapse will grow or wither? The story starts with the release of glutamate from the glutamatergic synaptic terminal. The binding of this glutamate to its two main inotropic receptors (AMPA, NMDA) not only leads to the influx of ions but also directly initiates various cell-signaling mechanisms independently of this influx, for example:

- The AMPAr activates the ERK-1 / ERK-2 MAP kinase cascade, which leads to the synthesis of more protein, including the AMPAr molecule. This cascade is also activated by BDNF via its receptor TrkB.
- The NMDAr activates eE2K kinase, which modulates the rate of synthesis of various dendritic proteins.
- NMDAr activation results in an increase in levels of the transcription factor c-fos in the limbic system and basal ganglia. NMDAr activation is also required for c-fos mRNA translation in a number of intracellular signaling systems.
- NMDAr activation directly affects axon–spine contacts in three ways:
- It causes postsynaptic neurons to secrete a protease into the synaptic cleft, which leads to the rapid hydrolysis of CAMs and thus to the loosening of the contact between the axon terminal and the spine.
- The lowered levels of calcium in the cleft following the NMDAr-mediated influx out of the synapse into the postsynaptic neuron destabilize interCAM contacts.

• Conversely, NMDAr activation causes *N*-cadherin to dimerize and so to become more resistant to proteolysis. This causes the axon terminal to stick closer to its spine.

• Metabotropic glutamate receptors initiate a number of cascades that involve cyclic nucleotides. These raise the level of protein synthesis in the spine via phosphokinase activation.

The influx of calcium into the postsynaptic neuron that is triggered by glutamate binding to the NMDAr has multiple effects. For example, it activates PLA2 and thus PGHs and NOS (when present[1]). This process leads to the release of a number of volume-signaling molecules: AA, ROS (in particular, hydrogen peroxide), NO, and various prostaglandins. Like the messengers sent forth by Lars Porsena in Longfellow's poem, these ride east, west, south, and north to summon the array of further biochemical reactions detailed above. The increased intracellular calcium levels also destabilize the link between the NMDAr subunit NR1 and alpha-actinin-2 as well as destabilizing actin itself. These two reactions will lead to changes in spine and dendrite morphology. The inflow of calcium is compartmentalized into at least two separate channels. Recent excellent reviews of the compartmentalization of calcium in the dendritic spine have been given by Yuste et al. (2000) and Sabatini et al. (2001).

The other main players in this process are the many cell adhesion molecules, scaffolding proteins, cytoskeletal proteins, nuclear transcription factors, neurotropins, and their attendant enzymes that form many cascades interacting with each other, with receptors and other cell surface proteins, and with the nucleus in the many complex ways described in this book. Most of this coordinated activity is designed to promote the efficiency of existing synapses and to recruit new ones, or, alternatively, to downregulate and remove unsuccessful synapses.

A successful synapse sets in motion processes that fine tune (upregulate) its existing receptors and channels, and deliver new membrane, cytoskeletal, and scaffolding proteins, enzymes, and signaling systems to the building site. It also arranges for the synthesis of this new material at nuclear and local levels. Most of this biochemical activity does not take place in the cytoplasmic soup, but on the production lines of an interwoven series of precisely engineered factories.[2]

Moreover, these factories are not set on a stable platform, but themselves are caught up in the ceaseless dynamic process of endocytosis, biochemical processing in the endosome system, and recycling back to the membrane. The staggering complexity of the system makes one sympathetic to the remark made by Turrigano (2000, p. 5): "The neuroscientist studying synaptic plasticity may come to feel like Hercules fighting the many-headed Hydra: for each question dispatched, two more spring up to take its place."

Appendix A

mRNAs Whose Level Is Altered Following Glutamate Receptor Stimulation

The paper by Hevroni et al. (1998) lists a large number of mRNAs whose level in the postsynaptic cell is altered after stimulation of the glutamate receptor. Minor players from their list not so far considered (or only briefly mentioned) in this book are the following:

Activin

This is a member of the transforming growth factor-β superfamily, which regulates various physiological functions in the brain. It increases the expression of voltage-dependent calcium channels through the activation of a tyrosine kinase and protein kinase C (Fukuhara et al. 1997; Iwahori et al. 1997). It increases luteinizing hormone release in the hypothalamus (MacConell et al. 1998). It also cooperates with fibroblast growth factor 2 in producing a rapid increase in tyrosine hydroxylase levels in cultured ventricular zone progenitors in the developing brain (Daadi et al. 1998). It is interesting that fibroblast growth factor is another neurotropin like nerve growth factor (NGF) whose mitogenic activity not only is maintained after it is endocytosed but actually depends on this endocytosis (Grieb and Burgess 2000).

Monocyte chemoattractant protein-1 (MCP-1) is a chemokine, that is, a type of cytokine whose function includes the correct positioning of cells in tissues (Xia and Hyman 1999).

Insulin Growth Factors

Insulin growth factors (IGFs) have various actions in the brain. For example, IGF-1 acts on both neurons and oligodendrocytes to promote myelination (Dubois-Dalcq and Murray 2000). Induction of IGFs and IGF binding proteins follows brain injury in reaction to cytokines (Wood et al. 1995). IGFs may act as both paracrine and endocrine regulators of neuronal growth. IGF-1 increases the branching and total extent of both apical and basal dendrites of pyramidal cells in rat somatosensory cortex, whereas brain-derived neurotropic factor

(BDNF) and neurotrophin-3 (NT-3) enhanced only basal dendritic development (Niblock et al. 2000).

Insulin itself can be either neurotoxic or neuroprotective in the same cell system by way of different cascades (Noh et al. 1999b). Whereas insulin attenuates neuronal apoptosis by activating phosphatidyl inositol-3 (PI-3) kinase in mouse cortical cell culture, it paradoxically induces neuronal necrosis following a 48-hour exposure. Insulin neurotoxicity is mediated by tyrosine phosphorylation of the insulin receptor and of the IGF-1 receptor, with subsequent activation of PKC and is blocked by antioxidants, but not by caspase inhibitors. Insulin also increases levels of the N-methyl-D-aspartate receptor (NMDAr) subunit NR2A without much effect on NR1 or NR2B levels (Noh et al. 1999b).

Pigment epithelium-derived factor (PEDF) induces neuronal differentiation and protects against glutamate neurotoxicity. It specifically protects against hydrogen peroxide-induced apoptosis (Bilak et al. 1999; Cao et al. 1999).

S100beta is an astrocyte-derived cytokine that promotes neurite growth. It is overexpressed in Alzheimer's disease (Sheng et al. 2000). An increase of S100beta expression in the hippocampus is a feature of kainic acid-induced neurotoxicity (Bendotti et al. 2000) when it may be involved in the structural reorganization of the hippocampus.

Syndecan-2 is a cell-surface heparan sulfate proteoglycan that plays a critical role in spine development in that, in conjunction with PDZ domain scaffolding proteins, it organizes the transformation from the primitive long thin protrusions (filopodia) to mature spines with stubby and headed shapes (Ethell and Yamaguchi 1999).

Arc (or activity-related cytoskeletal-associated protein) is thought to play an important role in neuronal plasticity following activation of the NMDAr (Kunizuka et al. 1999). Arc is enriched in dendrites, where it associates with cytoskeletal proteins. Arc and Arc mRNA accumulate in dendrites at sites of recent activity. Blockade of Arc expression by a local infusion of antisense oligodeoxynucleotides into the hippocampus inhibits the maintenance of long-term potentiation (LTP) (but not its induction) and blocks the consolidation (but not the acquisition) of long-term memories (Guzowski et al. 2000). Arc together with the brain-specific protein, amida, may also modulate cell death in the brain (Irie et al. 2000).

SC2 is also known as axonin-1. It acts as a receptor for Nr-CAMs in modulating axonal growth and guidance (Lustig et al. 1999). The details of this complex system are described by Kunz et al. 1998.

Neurofilament-L (NF-L) is a major binding protein for protein phosphatase-1 (PP-1) and may target the functions of PP-1 in membranes and the cytoskeleton of neurons (Terry-Lorenzo et al. 2000).

ABGP is the CAM neurofascin.

Fibrillarin is a 34-kDa nucleolar protein that plays a role in ribosomal RNA processing (Pearson et al. 1999).

Hsc70 is a heatshock protein that, in conjunction with auxillin, is essential for the dissociation of clathrin from used vesicles (Greener et al. 2000). Hsc70 also interacts with cysteine string protein (CSP) in vesicular function (Stahl et al. 1999). In general, heatshock proteins are chaperone proteins that inhibit the aggregation of partly folded proteins and refold them (Ohtsuka and Suzuki 2000). They are involved in synaptic plasticity via action on local protein synthesis. They also combat oxidative stress by suppressing the accumulation of oxidatively damaged proteins (Ohtsuka and Suzuki 2000).

CSP is a chaperone protein for vesicle-associated membrane protein (VAMP), which has little secondary structure by itself and needs CSP in order to take up the required conformation (Chamberlain and Burgoyne 2000). The molecular mechanisms of exocytosis involve many protein–protein interactions that depend on proteins undergoing transition between various conformations in many of which chaperone proteins are involved. Another example of this is provided by the interaction between the proteins neuronal-Sec-1 (nSec-1) and syntaxin-1a in membrane fusion. X-ray crystallography has shown that nSec-1 has a horseshoe shape that wraps around the molecule of syntaxin and holds it in the new conformation required (Hanson 2000).

VGF is a polypeptide that is secreted by neurons and is abundant in the hypothalamus. It is regulated in the brain by electrical activity, injury, and the circadian clock. VGF mRNA is induced in the hypothalamic arcuate nucleus of fasting rats. It is thought, among other things, to play a critical role in energy homeostasis (Hahm et al. 1999).

Appendix B
Receptors That Are Endocytosed

Among the receptors relevant to central nervous system function that are known to be endocytosed are the following (this list is not exclusive):

- Muscarinic acetylcholine (Bernard et al. 1998; Sorensen et al. 1998; Szekeres et al. 1998)
- Beta-adrenergic (Cao et al. 1998; Hirasawa et al. 1998; Laporte et al. 1999)
- Dopamine (Dumartin et al. 1998)
- Serotonin (Hu et al. 1993)
- A3 Adenosine (Ferguson and Palmer 1999)
- Opioids (Ignatova et al. 1999; McConalogue et al. 1999)
- Receptors for various polypeptides and proteins (including hormones, growth factors, and cytokines) (Grimes et al. 1996; Koenig and Edwardson 1997; Jans and Hassan 1998; Petrou and Tashjian, 1998)
- Substance P (Trejo and Coughlin 1999; Böhm et al. 1997; McConalogue et al. 1999; Grady et al. 1996)
- Insulin (di Guglielmo et al. 1998)
- Epidermal growth factor (Vieira et al. 1996; Skarpen et al. 1998)
- Nerve growth factor (NGF) and TrkA (Grimes et al. 1996)
- Transferrin (Spiro et al. 1996)
- Ferritin (Hulet et al. 2000)
- Interleukin-2 (IL-2) (Subtil et al. 1996)
- Polymeric immunoglobulin A2 (Gibson et al. 1998)
- Alpha-amino-3-hydroxy-5-methyl-4-isoxazole proprionic acid (AMPA) receptors (Morales and Goda, 1999; Carroll et al. 1999). There are significant amounts of the AMPA subunit Glu-R2 inside the postsynaptic cell. The glutamate transporter molecule EAAC1 (excitatory amino acid carrier 1) also occurs inside spines and dendritic shafts, giving evidence of endocytosis (He et al. 2000).
- Metabotropic glutamate (Doherty et al. 1999)

- The gonadotropin-releasing hormone receptor is internalized, but only exceptionally slowly. It is possible that the fact that this protein lacks an intracellular carboxyl terminal tail may account for this slowness (Vrecl et al. 2000).
- Integrins (Pierini et al. 2000)
- High-density lipoproteins mediated by their receptors, cubilin and megallin (Hammad et al. 2000)
- Autocrine motility factor (a.k.a. neuroleukin and maturation factor), which is essential for cell motility (Le et al. 2000)

As well as

- Thrombin (Trejo and Coughlin 1999)
- Rabies virus (Lewis and Lentz 1999)
- Influenza virus (Roy et al. 2000)
- NA^+/K^+-ATPase (Chilbalin et al. 1998)
- Na^+/H^+ exchanger (Kurashima et al. 1998)

Notes

Chapter 1

1. Proteosomes are large (~700-kDa) complexes composed of twenty-eight alpha and beta subunits arranged in four rings. Each ring has seven alpha and seven beta subunits. Proteosomes catabolize in particular oxidized proteins (Keller et al. 2000).
2. Lysosomes contain more than fifty acid-dependent hydrolases (e.g., proteases, lipases, and glycosidases) (Dell'angelica et al. 2000).
3. Among the damaging effects of superoxide are direct oxidation of low molecular weight reductants, inactivation of certain enzymes, combination with NO to produce $ONOO^-$, and release of free iron from [4 Fe-4 S] clusters of dehydratases (Liochev and Fridovich 1999). However, Thiels et al. (2000) have shown that mice that over-express SOD show lowered levels of LTP in CA1 in the hippocampus. Extracellular-superoxide dismutase (EC-SOD) transgenic mice show defects in long-term but not short-term memory for fear conditioning. These authors suggest that superoxide may not just be an unwanted toxin, but may be a signaling molecule in its own right.
4. Different dopamine postsynaptic cascades have been shown to be linked to different behaviors (Undie 2000). The cAMP pathway is linked to the jaw movements and the diacylglycerol pathway to the stereotyped behaviors and catalepsy characteristic of dopamine overstimulation.
5. The effect of inhibition and promotion of endocytosis on new synapse formation depends on microanatomical considerations. Inhibition of endocytosis on one spine will leave more membrane on that spine available for local growth at the cost of making less membrane available for new synaptic growth in adjacent regions.

Chapter 2

1. Recent data suggest that late endosomes and lysosomes fuse to form a hybrid organelle. Dense-core lysosomes are storage granules for acid hydrolases, which, in the hybrid organelle, digest endocytosed macromolecules. Rab, NSF, and Tether and SNARE proteins take part in this process (Luzio et al. 2000).
2. Other proteins involved in endocytosis include intersectin, which contains an Eps-15 homology and associates with clathrin (Hussain et al. 1999); annexins that bind phospholipids and are involved in membrane dynamics (Kobayashi et al. 1998); rababtins and growth-associated protein-43 (GAP-43), which mediate membrane fusion by interaction with calmodulin (Neve et al. 1998, N-ethyl malemide (NEM)-sensitive factor, which is required for the many membrane fusions involved in endocytosis (Lledo et al. 1999); receptor tyrosine kinases (rTKs) such as ErbB and ERK-1/2, which act as receptors for EGF and TGF-9; and endosomal cytosolic coat protein-1 (COP-1), proteins, which are involved in movements from early to late endosomes (Ceresa and

Schmid 2000; Gu and Gruenberg 1999). COP proteins include clathrin, COP-1 (a.k.a. coatomer), and COP-2.

3. Triglyceride surface lipoproteins, such as apolipoproteins C, E, and S, are also endocytosed and partly recycled and partly degraded (Heeren et al. 1999). ATP applied extracellularly doubles the amount of membrane trafficking in brown adipocytes (Pappone and Lee 1996). ATP acting on its P1 receptor is an inhibitory neurotransmitter, whereas adenosine acting on its P2 receptor acts as an excitatory neurotransmitter. ATP is colocalized and coreleased with ACh, norepinephrine (NE), glutamate, GABA, and neuropeptide Y at synapses (Williams and Jarvis 2000).

Chapter 3

1. Wnt-7a unbundles microtubules in mossy fiber axons as does the antimanic drug, lithium. There are two lithium-sensitive pathways in the brain: glycogen synthase kinase-3β (GSK-3β) and inositol(1,4,5)-triphosphate. The former is on the post-Wnt pathway, Wnt → frizzled protein receptor → dishevelled protein → GSK-3β (Williams and Harwood 2000).

2. There are three MAPK signaling pathways:

   - extracellular signal-regulated kinase
   - p38 kinase
   - c-Jun NH_2-terminal kinase (JNK)

   The last two are also known as stress-activated protein kinases (SAPK) (Cuenda 2000). For a recent review of the MAPK system, see Sweatt (2001).

3. A PDZ domain has about ninety amino acid residues with a glycine-leucine-glycine-phenylalanine repeat motif. Most PDZ domains contain two alpha helices and six beta sheets (Garner et al. 2000).

4. For more on clustering molecules, see the discussion on Homer in section 3.2.

Chapter 4

1. The membrane-permeable form of cAMP itself induces an increase in the number of synaptic boutons and the amount of membrane recycling in hippocampal neurons in tissue culture (Ma et al. 1999). Both isoforms II and IV of adenylate cyclase are linked to synaptic plasticity (Chern 2000).

Chapter 5

1. The Notch receptor is a single transmembrane protein whose extracellular domain recognizes DSL (Delta/Serrate/Lag2)-type cell-bound ligands. Activation leads to at least three proteolytic cleavages of its molecule and the release of its intracellular domain, which is trafficked to the nucleus and there modulates gene expression. Notch plays an active role in, among other things, glial cell differentiation and modulation of the length and organization of neuronal processes (Frisén and Lendahl 2001).

Chapter 6

1. There is compartmentalization of calcium pathways inside neurons. For example, the pathways mediating the effect of calcium on gene expression are different from those that mediate synaptic plasticity (Sattler et al. 1998).

Abbreviations and Acronyms

AA	arachidonic acid
ABGP	another name for the cell adhesion molecule neurofascin
Ac	adenylate cyclase
ACh	acetylcholine
ACSs	adenylate cyclases
ADAM	A disintegrin and a metalloprotein domain
ADP	adenosine diphosphate
AKAP	A kinase anchoring protein
Ala	alanine
AMP	adenosine monophosphate
AMPA	alpha-amino-3-hydroxy-5-methyl-4-isoxazoleproprionic acid
AMPArs	AMPA receptors
AP-1	clathrin adaptor protein-1; also 2
ApoE	apoenzyme E
APP	amyloid precursor protein
Arc	activity-related cytoskeletal-associated protein
ATP	adenosine triphosphate
ATPase	adenosine triphosphatase
BDNF	brain-derived neurotropic factor
BP-AP	back-propagated action potential
CAM	cell adhesion molecule
CaMKII	alpha Ca^{2+}-calmodulin-dependent kinase II
cAMP	cyclic adenosine monophosphate
CAMs	cell adhesion molecules
CarG	an actin gene enhancer
Cas	P130 CT-10 regulated kinase (Crk)-associated substrate
CASK	a membrane modular adapter protein kinase containing signaling modules CaM kinase and SH3

CAT	catalase
Cb1	an ubiquitine-protein ligase
CCE	capacitative calcium entry
CDK	cyclin-dependent kinase
cGMP	cyclic guanosine 5′-monophosphate
CLIP	cytoplasmic linker protein
CNS	central nervous system
COMT	catecholamine-O-methyl transferase
Con A	concanavalin A
COP	cytosolic coat protein
COX	cyclo-oxygenase
CRASH	a genetic disorder: corpus callosum agenesis, retardation, adducted thumbs, spastic pariesis, hydrocephalus
CREB	cyclic adenosine monophosphate response element-binding protein
CRMP	collapsin-response mediator protein
CSF	cerebrospinal fluid
CSP	cysteine-string protein
Cys	cysteine
D1R	dopamine D1 receptor; also D2R
Dab	disabled protein
DAG	diacylglycerol
DAQ	dopamine o-quinone
DARPP	dopamine and adenosine 3′-5′-monophosphate regulated phosphoprotein
DAT	dopamine transporter
DCC	deleted in colon cancer (netrin receptor)
DIGs	detergent-insoluble glycolipid- and cholesterol-enriched membrane rafts
DPF/W	arginine/proline/phenylalanine or tryptophan domain protein
DUBs	deubiquitinylating enzymes
EAAC1	excitatory amino acid carrier protein
EBP50	(a.k.a. NHERF, see entry) ezrin-radizin-moesin-binding phosphoprotein
EC-SOD	extracellular superoxide dismutase
eEF-2	eukaryotic elongation factor-2
EGF	epidermal growth factor

EGFr	epidermal growth factor receptor
EPSC	excitatory postsynaptic currents
ER	endoplasmic reticulum
ERK	extracellular signal-regulated kinase
FAK	focal adhesion kinase
FGF	fibroblast growth factor
FMRP	fragile X mental retardation protein
FRZ	frizzled
G1T1	a guanosine triphosphatase-activating protein for the adenosine diphosphate ribosylation factor of small GTP-binding proteins
GABA	gamma-aminobutyric acid
GABARAP	GABA receptor-associated protein
GAP	growth-associated protein
GDP	guanosine diphosphate
GEFs	guanine nucleotide exchange factors
GKAP	guanylate kinase domain-associated protein
Glu	glutamate
Glurs	glutamate receptors
GluTs	glutamate transporters
Gly	glycine
GPCr	G-protein-coupled receptor
GRASPs	glutamate receptor-interactive protein-associated proteins
GRIP	glutamate receptor-interactive protein
GR	G-protein coupled
GRP	glucose-regulated protein
GrKs	G-protein-coupled receptor kinase
GSHpx	glutathione peroxidase
GSK	glycogen synthase kinase
GTP	guanosine 5'-triphosphate
Hiw	product of the gene highwire
Hsc	heatshock cognate protein
5HT	5-hydroxytryptamine
ICE	isoconverting enzyme (a.k.a. caspase-1)
IGFs	insulin growth factors
Ig	immunoglobulin

Ig-SF	immunoglobulin superfamily
IL-6	interleukin-6
JNK	c-jun NH_2-terminal kinase
LGN	lateral geniculate nucleus
LH-RH	luteinizing hormone–releasing hormone
LTD	long-term depression
LTP	long-term potentiation
MAG	myelin-associated glycoprotein
MAGUK	membrane-associated guanylate kinase
MAP	microtubule-associated protein
MAPK	mitogen-associated protein kinase
MARCKs	myristolated alanine-rich C kinases
MAT	methionine-adenine transferase
MCP-1	monocyte chemoattractant protein-1
mGlurs	metabotropic glutamate receptors
mhip1R	a close relative of hip, the huntingtin interactive protein
mIPSs	miniature inhibitory postsynaptic potentials
MNC	magnocellular neuroendocrine cells
MNK	menkis protein
MSPs	miniature synaptic potentials
NAD	nicotinamide adenine dinucleotide
NADPH	reduced nicotinamide adenine dinucleotide phosphate
NCAM	neural cell adhesion molecule
NE	norepinephrine
NEM	N-ethyl maleimide
NF	neurofilament
NF-κB	nuclear factor κB
NGF	nerve growth factor
NGFr	nerve growth factor receptor
NHERF	Na^+/H^+ exchanger regulatory factor
NM	neuromodulator
NMDA	N-methyl-D-aspartate
NMDArs	N-methyl-D-apartate receptors
NO	nitric oxide
NOS	nitric oxide synthase
NR-1	units of NMDA receptor; also NR-2
Nr-CAM	Ng-CAM-related cell adhesion molecule

NSF	*N*-ethylamide-sensitive fusion protein
N-sec-1	a membrane fusion protein a.k.a. munc 18.1 and rbSec1
NT	neurotransmitter
NT-1	neurotropin-1; also 2 to 4
PDGF	platelet-derived growth factor
PDZ	postsynaptic density 95/Disc-large/ZO-1
PEDF	pigment epithelium-derived factor
PGE2	prostaglandin E2
PGH	prostaglandin H
PI-3K	phosphatidylinositol-3 kinase
PKA	cyclic adenosine monophosphate-dependent protein kinase A; also C
PLA	phospholipase A; also B, C, D and 1, 2
POB1	partner for ral-BP1
PP-1	protein phosphatase-1
PRD	proline-rich domain
Pro	Proline
PrPC	prion protein C
PSD	postsynaptic density
PYK2	proline rich tyrosine kinase 2
RAFTK	related adhesion focal tyrosine kinase
RAGE	receptor for advanced glycation end products
RNS	reactive nitrogen species
ROCK	rho-associated kinase
ROS	reactive oxygen species
rTK	receptor tyrosine kinase
SAP	synapse-associated kinase
SAPK	stress-activated protein kinases
Ser	serine
SH3	scr homology region 3
Shank	a family of synaptic proteins
SHMT	serine hydroxy methyl transferase
SNAP	stage-specific neurite-associated protein
SNARE	soluble *N*-ethylamide-sensitive fusion protein attachment protein receptor
SOD	superoxide dismutase
SON	supraoptic nucleus

Src	a family of protein kinases activated by ROS
TAG	transiently expressed axonal surface glycoprotein
Tbr-1	T-box transcription factor
TGF	transforming growth factor
Thr	threonine
Trk A	a receptor for nerve growth factor
Trp	tryptamine
TUC	the TOAD/ulip/CRMP family of growth cone proteins
Tyr	tyrosine
TyrK	tyrosine kinase
TyrKA	tyrosine kinase A
VAMP	vesicle-associated membrane protein (a.k.a. synapto-brevin)
VASE	variable alternative spliced exon
VASP	vasodilator-stimulated phosphoprotein
VIP	vasoactive intestinal peptide
VLDL	very low-density lipoprotein
VPL	ventroposterolateral
Wnt	"wingless" morphogen

References

Abe K, Saito H. (2000) The mitogen-activated protein kinase cascade mediates neurotrophic effect of epidermal growth factor in cultured rat hippocampal neurons. *Neurosci. Lett.* 282, 89–92.

Adamec RE, Burton P, Shallow T, Budgell J. (1999) NMDA receptors mediate lasting increases in anxiety-like behavior produced by the stress of predator exposure—implications for anxiety associated with posttraumatic stress disorder. *Physiol. Behav.* 65, 723–737.

Agostinho P, Duarte CB, Oliveria CR. (1997) Impairment of excitatory amino acid transporter activity by oxidative stress conditions in retinal cells: effect of antioxidants. *FASEB J.*, 11, 154–163.

Aizenman E, Lipton SA, Loring RH. (1989) Selective modulation of NMDA responses by reduction and oxidation. *Neuron* 2, 1257-1263.

Albright TD, Jessell TM, Kandel ER, Posner MI. (2000) Neural science: a century of progress and the mysteries that remain. *Neuron* 25, Suppl. S1–55.

Allen PB, Ouimet CC, Greengard P. (1997) Spinophilin, a novel protein phosphatase 1 binding protein localized to dendritic spines. *PNAS USA* 94, 9956–9961.

Allison DW, Chervin AS, Gelfand VI, Craig AM. (2000) Postsynaptic scaffolds of excitatory and inhibitory synapses in hippocampal neurons: maintenance of core components independent of actin filaments and microtubules. *J. Neurosci.* 20, 4545–4554.

Almeida A, Heales SJR, Bolaños JP, Medina JM. (1998) Glutamate neurotoxicity is associated with nitric oxide-mediated mitochondrial dysfunction and glutathione depletion. *Brain Res.* 790, 209–216.

Angelucci F, Aloe L, Vasquez PJ, Mathé AA. (2000) Mapping the differences in the brain concentration of brain-derived neurotropic factor (BDNF) and nerve growth factor (NGF) in an animal model of depression. *Neuroreport* 11, 1369–1373.

Angelucci F, Mathé AA, Aloe L. (2000) Brain-derived neurotrophic factor and tyrosine kinase receptor TrkB in rat brain are significantly altered after haloperidol and resperidone administration. *J. Neurosci. Res.* 60, 783–794.

Aoki C, Rhee J, Lubin M, Dawson TM. (1997) NMDA-R1 subunit of the cerebral cortex co-localizes with neuronal nitric oxide synthase at pre- and postsynaptic sites and in spines. *Brain Res.* 750, 25–40.

Arbuthnott GW, Ingham CA, Wickens JR. (1998) Modulation by dopamine of cat corticostriatal input. *Adv. Pharmacol.* 42, 733–736.

Artigiani S, Comoglio PM, Tamagnone L. (1999) Plexins, semaphorins, and scatter factor receptors: a common route for cell guidance signals? *Iubmb Life* 48, 477–482.

Asztely F, Erdemli G, Kullmann DM. (1997) Extrasynaptic glutamate spillover in the hippocampus: dependence on temperature and the role of active glutamate uptake. *Neuron* 18, 281–293.

Asztely F, Kokaia M, Olofsdotter I, Örtegen U, Lindvall O. (2000) Afferent-specific modulation of short-term synaptic plasticity by neurotrophins in dentate gyrus. *Eur. J. Pharmacol.* 12, 662–669.

Atkins CM, Sweatt JD. (1999) Reactive oxygen species mediate activity-dependent neuron-glia signaling in output fibers of the hippocampus. *J. Neurosci.* 19, 7241–7248.

Atlante A, Calissano P, Bobba A, Giannattasio S, Marra E, Passarella S. (2001) Glutamate neurotoxcity, oxidative stress and mitochondria. *FEBS Lett.* 497, 1–5.

Aubert I, Ridet JL, Schachner M, Rougon M, Gage FH. (1998) Expression of L1 and PSA during sprouting and regeneration in the adult hippocampal formation. *J. Comp. Neurol.* 399, 1–19.

Baez S, Segura-Aguilar J, Widersten M, Johansson AS, Mannervik, B. (1997) Glutathione transferases catalyse the detoxication of oxidized metabolites (*o*-quinones) of catecholamines and may serve as an antioxidant system preventing degenerative cellular processes. *Biochem. J.* 324, 25–28.

Bahr BA. (2000) Integrin-type signaling has a distinct influence on NMDA-induced cytoskeletal disassembly. *J. Neurosci. Res.* 59, 827–832.

Bains JS, Ferguson AV. (1997) Nitric oxide regulates NMDA-driven GABAergic inputs to type I neurones of the rat paraventricular nucleus. *J. Physiol.* 499, 733–746.

Baird DH, Trenkner E, Mason CA. (1996) Arrest of afferent axon extension by target neurons in vitro is regulated by the NMDA receptor. *J. Neurosci.* 16, 2642–2648.

Balschun D, Schneider H, Pitossi F, del Rey A, Besedovsky HO. (1998) The cytokine interleukin-1β has a key function in the maintenance of hippocampal long-term potentiation. *Eur. J. Neurosci.* 10, S10–S20.

Batchelor AM, Garthwaite J. (1997) Frequency detection and temporally dispersed synaptic signal association through a metabotropic glutamate receptor pathway. *Nature* 385, 74–77.

Beattie EC, Zhou J, Grimes ML. Bunnett NW, Howe CL, Mobley WC. (1996) A signaling endosome hypothesis to explain NGF actions: potential implications for neurodegeneration. *CSHSQB* 61, 389–406.

Beattie EC, Carroll RC, Yu X, Morishita W, Yasuda H, von Zastrow M, Malenka RC. (2000a) Regulation of AMPA receptor endocytosis by a signaling mechanism shared with LTD. *Nature Neuroscience* 3, 1291–1300.

Beattie EC, Howe CL, Wilde A, Brodsky FM, Mobley WC. (2000b) NGF signals through TrkA to increase clathrin at the plasma membrane and enhance clathrin-mediated membrane trafficking. *J. Neurosci.* 20, 7325–7333.

Behrens J. (2000) Cross-regulation of the Wnt signalling pathway: a role of MAP kinases. *J. Cell Sci.* 113, 911–919.

Bendotti C, Guglielmetti F, Tortarolo M, Samanin R, Hirst WD. (2000) Differential expression of S100beta and glial fibrillary acidic protein in the hippocampus after kainic acid-induced lesions and mossy fiber sprouting in the adult rat. *Exp. Neurol.* 161, 317–329.

Ben-Shachar D, Zuk R, Glinka Y. (1995) Dopamine neurotoxicity: inhibition of mitochondrial respiration. *J. Neurochem.* 64, 718–723.

Berg MM, Krafft GA, Klein WL. (1997) Rapid impact of beta-amyloid on paxillin in a neural cell line. *J. Neurosci. Res.* 50, 979–989.

Bergles DE, Jahr CE. (1997) Synaptic activation of glutamate transporters in hippocampal astrocytes. *Neuron* 19, 1297–1308.

Bergles DE, Diamond JS, Jahr CE. (1999) Clearance of glutamate inside the synapse and beyond. *Curr. Opin. Neurobiol.* 9, 293–298.

Berman SB, Hastings TG. (1999) Dopamine oxidation alters mitochondrial respiration and induces permeability transition in brain mitochondria: implications for Parkinson's disease. *J. Neurochem.* 73, 1127–1137.

Bernard V, Laribi O, Levey AI, Bloch B. (1998) Subcellular redistribution of m2 muscarinic acetylcholine receptors in striatal interneurons in vivo after acute cholinergic
stimulation. *J. Neurosci.* 18, 10,207–10,218.

Bhargavi V, Chari VB, Singh SS. (1998) Phosphatidylinositol 3-kinase binds to profilin
through the p85 alpha subunit and regulates cytoskeletal assembly. *Biochem. Mol.
Biol. Internat.* 46, 241–248.

Bilak MM, Corse AM, Bilak SR, Lehar M, Tombran-Tink J, Kuncl RW. (1999) Pigment
epithelium-derived factor (PEDF) protects motor neurons from chronic gluta-
mate-mediated neurodegeneration. *J. Neurpathol. Exp. Neurol.* 58, 719–728.

Black IB. (1999) Trophic regulation of synaptic plasticity. *J. Neurobiol.* 41, 108–118.

Blake DR, Winyard P, Lunec J, Williams A, Good PA, Crewes SJ, Gutteridge JMC, Rowley
D, Halliwell B, Cornish A, Hider RC. (1985) Cellular and ocular toxicity induced
by desferrioxamine. *Quart. J. Med.* 56, 345–355.

Blennow K, Davidsson P, Gottfries C-G, Ekman R, Heilig M. (1996) Synaptic degeneration in thalamus in schizophrenia. *Lancet* 348, 692–693.

Blennow K, Bogdanovic N, Gottfries C-G, Davidsson P. (2000) The growth-associated
protein GAP-43 is increased in the hippocampus and in the gyrus cinguli in
schizophrenia. *J. Mol. Neurosci.* 13, 101–109.

Bloch-Gallego E, Ezan F, Tessier-Lavigne M, Sotelo C. (1999) Floor plate and netrin-1
are involved in the migration and survival of inferior olivary neuurons. *J. Neurosci.*
19, 4407–4420.

Böhm SK, Khitin LM, Smeekens SP, Grady EF, Payan DG, Bunnett NW. (1997) Identification of potential tyrosine-containing endocytic motifs in the carboxyl-tail and sev-
enth transmembrane domain of the neurokinin 1 receptor. *J. Biol. Chem.* 272,
2363–2372.

Boldyrev AA, Stvolinsky SL, Tyulina OV, Koshelev VB, Hori N, Carpenter DO. (1997)
Biochemical and physiological evidence that carnosine is an endogenous neuro-
protector against free radicals. *Cell. Mol. Neurobiol.* 17, 259–271.

Boldyrev A, Song R, Lawrence D, Carpenter DO. (1999) Carnosine protects aganst excitotoxic cell death independently of effects on reactive oxygen species. *Neuroscience*
94, 571–577.

Bonfoco E, Leist M, Zhivotosky B, Orrenius S, Lipton SA, Nicotera P. (1996) Cytoskeletal
breakdown and apoptosis elicited by NO donors in cerebellar granule cells require
NMDA receptor activation. *J. Neurochem.* 67, 2484–2493.

Bothwell M, Giniger E. (2000) Alzheimer's disease: neurodevelopment converges with
neurodegeneration. *Cell* 102, 271–273.

Bottomley MJ, Surdo PL, Driscoll PC. (1999) Endocytosis: how dynamin gets vesicles
PHree! *Curr. Biol.* 9, R301–R304.

Bovolenta P, Fernaud-Espinosa I. (2000) Nervous system proteoglycans as modulators
of neurite outgrowth. *Prog. Neurobiol.* 61, 113–132.

Brand A, Leibfritz D, Richter-Landsberg C. (1999) Oxidative stress-induced metabolic
alterations in rat brain astrocytes studied by multinuclear NMR spectroscopy. *J.
Neurosci. Res.* 58, 576–585.

Brandner C, Vantini G, Schenk F. (2000) Enhanced visuospatial memory following intra-
cerebroventricular administration of nerve growth factor. *Neurobiol. Learn. Memory*
73, 49–67.

Breder CD, Smith WL, Raz A, Masferrer J, Seibert K, Needleman P, Saper CB. (1992)
Distribution and characterization of cyclooxygenase immunoreactivity in the
ovine brain. *J. Comp. Neurol.* 322, 409–438.

Bretscher MS, Aguado-Velasco C. (1998) Membrane traffic during cell locomotion. *Curr.
Opin. Cell Biol.* 10, 537–541.

Breuer W, Epsztejn S, Cabantchik B. (1995) Iron acquired from transferrin by K562 cells is delivered to a cytolasmic pool of chelatable iron (II). *J. Biol. Chem.* 270, 24,209–24,215.

Britto JM, Tannahill D, Keynes RJ. (2000) Life, death and Sonic hedgehog. *Bioessays* 22, 499–502.

Brodin L, Löw P, Shupliakov O. (2000) Sequential steps in clathrin-mediated synaptic vesicle endocytosis. *Curr. Opin. Neurobiol.* 10, 312–320.

Broide RS, Leslie FM. (1999) The alpha-7 nicotinic acetylcholine receptor in neuronal plasticity. *Mol. Neurobiol.* 20, 1–16.

Brown DR. (2000) Neuronal release of vasoactive intestinal peptide is important to astrocyte protection of neurons from glutamate toxicity. *Mol. Cell. Neurosci.* 15, 465–475.

Brown GC. (1997) Nitric oxide inhibition of cytochrome oxidase and mitochondrial respiration: implications for inflammatory, neurodegenerative and ischaemic pathology. *Mol. Cell. Biochem.* 174, 189–192.

Brown QB, Bahr BA. (2000) Phosphorylation of the ERK1/ERK2 MAP kinase can be influenced by AMPA-type glutamate receptors and their positive modulation in hippocampus. *J. Neurochem.* 74S, S20.

Burack WR, Shaw AS. (2000) Signal transduction: hanging on a scaffold. *Curr. Opin. Cell Bio.* 12, 211–216.

Burden-Gulley SM, Lemmon V. (1996) L1, N-cadherin, and laminin induce distinct distribution patterns of cytoskeletal elements in growth cones. *Cell Mot. Cytoskel.* 35, 1–23.

Burrone J, Murthy VN. (2001) Synaptic plasticity: rush hour traffic in the AMPA lanes. *Curr. Biol.* 11, R244–R277.

Bush A, Huang X, Cherny RA, Lynch T, Goldstein LE, Atwood CS, Moir RD, Li Q-X, Cabelli DE, Multhaup G, Roher AE, Tanzi RE, Masters CL. (2000) Evidence that β-amyloid in Alzheimer's disease represents a corrupted antioxidant. *Neuropsychopharmacology* 23, S52.

Cadet JL, Kahler LA. (1994) Free radical mechanisms in schizophrenia and tardive dyskinesia. *Neurosci. Biobehav. Rev.* 18, 457–467.

Calizo LH, Flanagan-Cato LM. (2000) Estrogen selectively regulates spine density within the dendritic arbor of rat ventromedial hypothalamic neurons. *J. Neurosci.* 20, 1589–1596.

Cambonie G, Laplanche L, Kamenka J-M, Barbanel G. (2000) N-Methyl-D-aspartate but not glutamate induces the release of hydroxyl radicals in the neonatal rat; modulation by group I metabotropic glutamate receptors. *J. Neurosci. Res.* 62, 84–90.

Camel JE, Withers GS, Greenough WT. (1986) Persistence of visual cortex dendritic alterations induced by postweaning exposure to a "superenriched" environment in rats. *Behav. Neurosci.* 100, 810–813.

Cao TT, Mays RW, von Zastrow M. (1998) Regulated endocytosis of G-protein-coupled receptors by a biochemically and functionally distinct subpopulation of clathrin-coated pits. *J. Biol. Chem.* 273, 24,592–24,602.

Cao W, Tombran-Tink J, Chen W, Mrazek D, Elias R, McGinnis JF. (1999) Pigment epithelium-derived factor protects cultured retinal neurons against hydrogen peroxide-induced cell death. *J. Neurosci. Res.* 57, 789–800.

Cao X, Barlowe C. (2000) Asymmetric requirement for a Rab GTPase and SNARE proteins in fusion of COPII vesicles with acceptor membranes. *J. Cell Biol.* 149, 55–66.

Carlsson A, Waters N, Hansson LO. (1994) Neurotransmitter aberrations in schizophrenia: new findings. In: R. Fog, J. Gerlach and R. Hemmingsen (Eds.) *Schizophrenia. An Integrated View.* Copenhagen, Munksgard.

Carman CV, Benovic JL. (1998) G-protein-coupled receptors: turn-ons and turn-offs. *Curr. Opin. Neurobiol.* 8, 335–344.

Carroll RC, Beattie EC, Xia H, Lüscher C, Altschuler Y, Nicoll RA, Malenka RC, von Zastrow M. (1999) Dynamin-dependent endocytosis of ionotropic glutamate receptors, *PNAS USA* 96, 14112–14117.

Casadio A, Martin KC, Giustetto M, Zhu H, Chen M, Bartsch D, Bailey CH, Kandel ER. (1999) A transient, neuron-wide form of CREB-mediated long-term facilitation can be stabilized at specific synapses by local protein synthesis. *Cell* 99, 221–237.

Cataldo AM, Barnett JL, Pieroni C, Nixon RA. (1997) Increased neuronal endocytosis and protease delivery to early endosomes in sporadic Alzheimer's disease: neuropathologic evidence for a mechanism of increased bets-amyloidogenesis. *J. Neurosci.* 17, 6142–6151.

Centonze D, Gubellini P, Bernardi G, Calabresi P. (1999) Permissive role of interneurons in corticostriatal synaptic plasticity. *Br. Res. Rev.* 31, 1–5.

Ceresa BP, Schmid SL. (2000) Regulation of signal transduction by endocytosis. *Curr. Opin. Cell Biol.* 12, 204–210.

Cerione RA. (2000) Cell signaling: a spider's web of architectural beauty and complexity. Review of "Signaling networks and cell cycle control." In *The Molecular Basis of Cancer and Other Diseases.* Gutkind FS. (Ed.) Totowa N.J.: Humana Press. In *Cell* 103, 555–556.

Cestra G, Castagnoli L, Dente L. Minenkova O, Petrelli A, Migone N, Hoffmüller U, Schneider-Mergener J, Cesareni G. (1999) The SH3 domains of endophilin and amphiphysin bind to the proline-rich region of synaptojanin 1 to distinct sites that display an unconventional binding specificity. *J. Biol. Chem.* 274, 32,001–32,007.

Chain DG, Schwartz JH, Hegde AN. (2000) Ubiquitin-mediated proteolysis in learning and memory. *Mol. Neurobiol.* 20, 125–142.

Chakraborti T, Ghosh SK, Michael JR, Batabyal SK, Chakraborti S. (1998) Targets of oxidative stress in cardiovascular system. *Mol. Cell. Biochem.* 187, 1–10.

Chamberlain LH, Burgoyne RD. (2000) Cysteine-string protein: the chaperone at the synapse. *J. Neurochem.* 74, 1781–1789.

Chen BT, Avshalumov MV, Rice ME. (2001) H_2O_2 is a novel, endogenous modulator of synaptic dopamine release. *J. Neurophysiol.* 85, 2468–2476.

Chen H, Fre S, Slepnev VI, Capua MR, Takei K, Butler MH, Di Fiore PP, De Camilli P. (1998) Epsin is an EH-domain-binding protein implicated in clathrin-mediated endocytosis. *Nature* 394, 793–797.

Chen H, Bagri A, Zupicich JA, Zou Y, Stoeckli E, Pleasure SJ, Lowenstein DH, Skarnes WC, Chédotal A, Skarnes WC, Tessier-Lavigne M. (2000) Neuropilin-2 regulates the development of selective cranial and sensory nerves and hippocampal mossy fiber projections. *Neuron,* 25, 43–56.

Chern Y. (2000) Regulation of adenylate cyclase in the central nervous system. *Cell Signal.* 12, 195–204.

Chibalin AV, Zierath JR, Katz AI, Berggren PO, Bertorello AM. (1998) Phosphatidylinositol 3-kinase-mediated endocytosis of renal Na^+,K^+-ATPase alpha subunit in response to dopamine. *Mol. Biol. Cell.* 9, 1209–1220.

Chin D, Means AR. (2000) Calmodulin: a prototypical calcium sensor. *TICB* 10, 322–328.

Chiueh CC, Rauhala P. (1999) The redox pathway of S-nitrosoglutathione, glutathione and nitric oxide in cell to neuron communications. *Free Rad. Res.* 31, 641–650.

Christian JL. (2000) BMP, Wnt and Hedgehog signals: how far can they go? *Curr. Opin. Cell Biol.* 12, 244–249.

Churchland PS, Sejnowski TE. (1992) *The Computational Brain.* Cambridge, MA. MIT Press.

Ciechanover A, Orian A, Schwartz AL. (2000) The ubiquitin-mediated proteolytic pathway: mode of action and clinical implications. *J. Clin. Biochem. Supp.* 34, 40–51.

Ciruela F, Soloviev MM. Chan WY, McIlhinney RA. (2000) Homer-1c / Vesl-IL modulates the cell surface targeting of metabotropic glutamate receptor type 1alpha: evidence for an anchoring function. *Mol. Cell. Neurosci.* 15, 36–50.

Clague MJ. (1998) Molecular aspects of the endocytic pathway. *Biochem. J.* 336, 271–282.

Claing A, Perry SJ, Achiriloaie M, Walker JK, Albanesi JP, Lefkowitz RJ, Premont RT. (2000) Multiple endocytic pathways of G-protein-coupled receptors delineated by GIT1 sensitivity. *PNAS USA* 97, 1119–1124.

Clément M-V, Ponton A, Pervaiz S. (1998) Apoptosis induced by hydrogen peroxide is mediated by decreased superoxide anion concentration and reduction of intracellular milieu. *FEBS Lett.* 440, 13–18.

Cogen J, Cohen-Cory S. (2000) Nitric oxide modulates retinal ganglion cell axon arbor remodeling *in vivo*. *J. Neurobiol.* 45, 120–133.

Colasanti M, Persichini T. (2000) Nitric oxide; an inhibitor of NF-κB / Rel system in glial cells. *Br. Res. Bull.* 52, 155–161.

Colledge M, Dean RA, Scott GK, Langeberg LK, Huganir RL, Scott JD. (2000) Targeting of PKA to glutamate receptors through a MAGUK-AKAP complex. *Neuron* 27, 107–119.

Collin C, Miyaguchi K, Segal M. (1997) Dendritic spine density and LTP induction in cultured hippocampal slices. *J. Neurophysiol.* 77, 1614–1623.

Comery TA, Stamoudis CY, Irwin SA, Greenough WT. (1996) Increased density of multiple-head dendritic spines on medium-sized spiny neurons of the striatum in rats reared in a complex environment. *Neurobiol. Learn. Mem.* 66, 93–96.

Conti F, Weinberg RJ. (1999) Shaping excitation at glutamatergic synapses. *TINS* 22, 451–458.

Corvera S, D'Arrigo A, Stenmark H. (1999) Phosphoinositides in membrane traffic. *Curr. Opin. Cell Biol.* 11, 460–465.

Cotman CW, Hailer NP, Pfister KK, Soltesz I, Schachner M. (1998) Cell adhesion molecules in neural plasticity and pathology: similar mechanisms, distinct organizations? *Prog. Neurobiol.* 55, 665–669.

Cotter D, Kerwin R, al-Sarraji S, Brion JP, Chadwich A, Lovestone S, Anderton B, Everall I. (1998) Abnormalities of Wnt signaling in schizophrenia—evidence of neurodevelopmental abnormality. *NeuroReport* 9, 1379–1383.

Couchman JR, Woods A. (1999) Syndecan-4 and integrins: combinatorial signaling in cell adhesion [comment]. *J. Cell Sci.* 112, 3415–3420.

Coussens CM, Teyler TJ.(1996) Protein kinase and phosphatase activity regulate the form of synaptic plasticity expressed. *Synapse* 24, 97–103.

Craig AM, Boudin H. (2001) Molecular heterogeneity of central synapses: afferent and target regulation. *Nature Neuroscience* 4, 269–278.

Critchley DR, Holt MR, Barry ST, Priddle H, Hemmings L, Norman J. (1999) Integrin mediated cell adhesion: the cytoskeletal connection. *Biochem. Soc. Symp.* 65, 79–99.

Cross RA, Carter NJ. (2000) Molecular motors. *Curr. Biol.* 10, R177–R179.

Csala M, Braun L, Mile V, Kardon T, Szarka A, Kupcsulik P, Mandl J, Bánhegyi G. (1999) Ascorbate-mediated electron transfer in protein thiol oxidation in the endoplasmic reticulum. *FEBS Lett.* 460, 539–543.

Cuajungco MP, Goldstein LE, Nunomura A, Smith MA, Lim JT, Atwood CS, Huang X, Farrag YW, Perry G, Bush AI. (2000) Evidence that the β-amyloid plaques of Alzheimer's disease represent the redox-silencing and entombment of Aβ by zinc. *J. Biol. Chem.* 275, 19,439–19,442.

Cuenda A. (2000) Mitogen-activated protein kinase kinase 4 (MKK4). *Int. J. Biochem. Cell Biol.* 32, 581–587.

Daadi M, Arcellana-Panlilio MY, Weiss S. (1998) Activin co-operates with fibroblast growth factor 2 to regulate tyrosine hydroxylase expression in the basal forebrain ventricular zone progenitors. *Neuroscience* 86, 867–880.

Dai J, Sheetz MP, Wan X, Morris CE. (1998) Membrane tension in swelling and shrinking molluscan neurons. *J. Neurosci.* 18, 6681–6692.

Damer CK, Creutz CE. (1994) Secretory and synaptic vesicle membrane proteins and their possible roles in regulating exocytosis. *Prog. Neurobiol.* 43, 511–536.

Daniels GM, Amara SG. (1999) Regulated trafficking of the human dopamine transporter. Clathrin-mediated internalization and lysosomal degradation in response to phorbol esters. *J. Biol. Chem.* 274, 35,794–35,801.

Davidsson P, Gottfries J, Bogdanovic N, Ekman R, Karlsson I, Gottfries C-G, Blennow K. (1999) The synaptic-vesicle-specific proteins rab3a and synaptophysin are reduced in thalamus and related cortical brain regions in schizophrenic brains. *Schiz. Res.* 40, 23–29.

Davies AM. (2000) Neurotrophins: neurotrophic modulation of neurite growth. *Curr. Biol.* 10, R198–R200.

Dawson VL. Dawson TM. (1996) Nitric oxide actions in neurochemistry. *Neurochem. Internat.* 79, 97–110.

Dawson VL, Dawson TM, London ED, Bredt DS, Snyder SH. (1991) Nitric oxide mediates glutamate neurotoxicity in primary cortical cultures. *PNAS USA* 88, 6368–6371.

De Ferrari GV, Inestrosa NC. (2000) Wnt signaling function in Alzheimer's disease. *Br. Res. Rev.* 33, 1–12.

Dell'Angelica EC, Mullins C, Caplan S, Bonifacino JS. (2000) Lysosome-related organelles. *FASEB J.* 14, 1265–1278.

DeLisi LE. (1997). Is schizophrenia a lifetime disorder of brain plasticity, growth and aging? *Schiz. Res.* 23, 119–129.

Denisova NA, Fisher D, Provost M, Joseph JA. (1999) The role of glutathione, membrane sphingomyelin, and its metabolites in oxidative stress-induced calcium "dysregulation" in PC12 cells. *Free Rad. Biol. Med.* 27, 1292–1301.

Desmond NL, Levy WB. (1998) Free postsynaptic densities in the hippocampus of the female rat. *NeuroReport* 9, 1975–1979.

de Wit R, Capello A, Boonstra J, Verkleij AJ, Post JA. (2000) Hydrogen peroxide inhibits epidermal growth factor receptor internalization in human fibroblasts. *Free Rad. Biol. Med.* 28, 28–38.

Diamond JS, Jahr CE. (1997) Transporters buffer synaptically released glutamate on a submillisecond time scale. *J. Neurosci.* 17, 4672–4687.

Dickie BGM, Holmes C, Greenfield SA. (1996) Neurotoxic and neurotrophic effects of chronic N-methyl-D-aspartate exposure upon mesencephalic dopaminergic neurons in organotypic culture. *Neuroscience* 72, 731–741.

Di Fiore PP, Gill GN. (1999) Endocytosis and mitogenic signaling. *Curr. Opin. Cell Biol.* 11, 483–488.

Di Guglielmo GM, Drake PG, Baass PC, Authier F, Posner BI, Bergeron JJM. (1998) Insulin receptor internalization and signaling. *Mol. Cell. Biochem.* 182, 59–63.

Dimitratos SD, Woods DF, Stathakis DG, Bryant PJ. (1999) Signaling pathways are focussed at specialized regions of the plasma membrane by scaffolding proteins of the MAGUK family. *Bioessays* 21, 912–921.

Dismukes K. (1977) New look at the aminergic neural system. *Nature* 269, 557.

Do KQ, Trabesinger AH, Kirsten-Krüger M, Lauer CJ, Dydak U, Hell D, Holsboer F, Boesiger P, Cuénod M. (2000) Schizophrenia: glutathione deficit in cerebrospinal fluid and prefrontal cortex in vivo. *Eur. J. Neurosci.* 12, 3721–3728.

Doherty P, Fazeli MS, Walsh FS. (1995) The neural cell adhesion molecule and synaptic plasticity. *J. Neurobiol.* 26, 437–446.

Doherty AJ, Coutinho V, Collingridge GL, Henley JM. (1999) Rapid internalization and surface expression of a functional, fluorescently tagged G-protein-coupled glutamate receptor. *Biochem. J.* 341, 415–422.

Dolcet X, Egea J, Soler RM, Martin-Zanca D, Comella JX. (1999) Activation of phosphatidylinositol 3-kinase, but not extracellular-regulated kinases, is necessary to mediate brain-derived neurotrophic factor-induced motoneuron survival. *J. Neurochem.* 73, 521–531.

Domínguez-Rosales JA, Mavi G, Levenson SM, Rojkind M. (2000) H_2O_2 is an important mediator of physiological and pathological healing processes. *Arch. Med. Res.* 31, 15–20.

Dong H, Zhang P, Song I, Petralia RS, Liao D, Huganir RL. (1999) Characterization of the glutamate receptor-interacting proteins GRIP1 and GRIP2. *J. Neurosci.* 19, 6930–6941.

Doria M, Salcini AE, Colombo E, Parslow TG, Pelicci PG, di Fiore P. (1999) The eps15 homology (EH) domain-based interaction between eps15 and hrb connects the molecular machinery of endocytosis to that of nucleocytosolic transport. *J. Cell Biol.* 147, 1379–1384.

Drescher U. (2000) Excitation at the synapse: Eph receptors team up with NMDA receptors. *Cell* 103, 1005–1008.

Dubois-Dalcq M, Murray R. (2000) Why are growth factors important in oligodendrocyte physiology? *Pathologie Biologie* 48, 80–86.

Dugan LL, Creedon DJ, Johnson EM Jr., Holtzman DM. (1997) Rapid suppression of free radical formation by nerve growth factor involves the mitogen-activated protein kinase pathway. *PNAS USA* 94, 4086–4091.

Dumartin B, Caillé I, Gonon F, Bloch B. (1998) Internalization of D1 dopamine receptor in striatal neurons in vivo as evidence of activation by dopamine agonists. *J. Neurosci.* 18, 1650–1661.

Eastwood SL, Harrison PJ. (1998) Hippocampal and cortical growth-associated protein-43 messenger RNA in schizophrenia. *Neuroscience* 86, 437–448.

Edgar PF. (2000) Comparative proteotome analysis. Tissue homogenate from normal human hippocampus subjected to two-dimensional gel electrophoresis and Coomassie blue protein staining. *Mol. Psychiat.* 5, 85–90.

Edgar PF, Douglas JE, Cooper GJ, Dean B, Kydd R, Faull RL. (2001) Comparative proteosome analysis of the hippocampus implicates chromosome 6q in schizophrenia. *Mol. Psychiatr.* 5, 85–90.

Edwardson JM, Szekeres PG. (1999) Endocytosis and recycling of muscarinic receptors. *Life Sci.* 64, 487–494.

Egberongbe YI, Gentleman SM, Falkai P, Bogerts B, Polak JM, Roberts GW. (1994) The distribution of nitric oxide synthase immunoreactivity in the human brain. *Neuroscience* 59, 561–578.

Ehlers MD. (1999) Synapse structure: glutamate receptors connected by the shanks. *Curr. Biol.* 9, R848–R850.

Ellis S, Mellor H. (2000) Regulation of endocytic traffic by rho family GTPases. *TICB* 10, 85–88.

Engert F, Bonhoeffer T. (1999) Dendritic spine changes associated with hippocampal long-term synaptic plasticity. *Nature* 399, 66–70.

Engqvist-Goldstein AE, Kessels MM, Chopra VS, Hayden MR, Drubin DG. (1999) The actin-binding protein of the Sla2/Huntingtin interacting protein 1 family is a novel component of clathrin-coated pits and vesicles. *J. Cell Biol.* 147, 1503–1518.

Ernst AF, Gallo G, Letourneau PC, McLoon SC. (2000) Stabilization of growing retinal axons by the combined signaling of nitric oxide and brain-derived neurotrophic factor. *J. Neurosci.* 20, 1458–1469.

Esch T, Lemmon V, Banker G. (1999) Local presentation of substrate molecules directs axon specification by cultured hippocampal neurons. *J. Neurosci.* 19, 6417–6426.

Estévez AG, Spear N, Manuel SM, Radi R, Henderson CE, Barbeito L, Beckman JS. (1998) Nitric oxide and superoxide contribute to motor neuron apoptosis induced by trophic factor deprivation. *J. Neurosci.* 18, 923–931.

Ethell IM, Yamaguchi Y. (1999) Cell surface heparan sulfate proteoglycan syndecan-2 induces the maturation of dendritic spines in rat hippocampal neurons. *J. Cell Biol.* 144, 575–586.

Ethell IM, Hagihara K, Miura Y, Irie F, Yamaguchi Y. (2000) Synbindin, A novel syndecan-2-binding protein in neuronal dendritic spines. *J. Cell Biol.* 151, 53–68.

Faivre-Sarrailh C, Falk J, Pollerberg E, Schachner M, Rougon G. (1999) NrCAM, cerebellar granule cell receptor for the neural adhesion molecule F3, displays an actin-dependent mobility in growth cones. *J. Cell Sci.* 112, 3015–3027.

Fanning AS and Anderson JM. (1999) Protein molecules as organizers of membrane structure. *Curr. Opin. Cell Biol.* 11, 432–439.

Farhadi HF, Mowla SJ, Petrecca K, Morris SJ, Seidah NG, Murphy RA. (2000) Neurotrophin-3 sorts to the constitutive secretory pathway of hippocampal neurons and is diverted to the regulatory secretory pathway by coexpression with brain-derived neurotrophic factor. *J. Neurosci.* 20, 4059–4068.

Feldman DE, Knudsen EI. (1998) Experience-dependent plasticity and the maturation of glutamatergic synapses. *Neuron* 20, 1067–1071.

Ferguson G, Palmer TM. (1999) Regulation of A3 adenosine receptor internalization by receptor phosphorylation. *Biochem. Soc. Trans.* 27, A115.

Ferguson SS, Caron MG. (1998) G protein-coupled receptor adaptation mechanisms. *Semin. Cell Dev. Biol.* 9, 119–127.

Fernández-Shaw C, Marina A, Cazorla P, Valdivieso F, Vázquez J. (1997) Anti-brain spectrin immunoreactivity in Alzheimer's disease: degradation of spectrin in an animal model of cholinergic degeneration. *J. Neuroimmunol.* 77, 91–98.

Fiala BA, Joyce JN, Greenough WT. (1978) Environmental complexity modulates growth of granule cell dendrites in developing but not adult hippocampus of rats. *Exp. Neurol.* 59, 372–383.

Fields RD, Itoh K. (1996) Neural cell adhesion molecules in activity-dependent development and synaptic plasticity. *TINS* 19, 473–480.

Figueiredo-Pereira ME, Cohen G. (1999) The ubiquitin/proteosome pathway: friend or foe in zinc-, cadmium-, and H_2O_2-induced neuronal oxidative stress. *Mol. Biol. Reports* 26, 65–69.

Fimia GM, Sassone-Corsi P. (2001) Cyclic AMP signaling. *J. Cell Sci.* 114, 1971–1972.

Firestein BL, Brenman JE, Aoki C, Sanchez-Perez AM, El-Husseini AE, Bredt DS. (1999) Cyprin: a cytosolic regulator of PSD-95 postsynaptic targeting. *Neuron* 24, 659–672.

Fischer M, Kaech S, Knutti D, Matus A. (1998) Rapid actin-based plasticity in dendritic spines. *Neuron* 20, 847–854.

Fischer M, Kaech S, Wagner U, Brinkhaus H, Matus A. (2000) Glutamate receptors regulate actin-based plasticity in dendritic spines. *Nature Neuroscience* 3, 887–894.

Folli F, Alvaro D, Gigliozzi A, Bassotti C, Kahn CR, Pontiroli AE, Capocaccia L, Jezequel AM, Benedetti A. (1997) Regulation of endocytic-transcytotic pathways and bile secretion by phosphatidylinositol 3-kinase in rats. *Gastroenterology* 113, 954–965.

Frantseva MV, Perez Velasquez JLP, Carlen PL. (1998) Changes in membrane and synaptic properties of the thalamocortical circuitry caused by hydrogen peroxide *J. Neurophysiol.* 80, 1317–1326.

Freigang J, Proba K, Leder L, Diederichs K, Sonderegger P, Welte W. (2000) The crystal structure of the ligand binding module of axonin-1/TAG-1 suggests a zipper mechanism for neural cell adhesion. *Cell* 101, 425–433.

Friedman HV, Bresler T, Garner CC, Ziv NE. (2000) Assembly of new individual excitatory synapses: time course and temporal order of synaptic molecule recruitment. *Neuron* 27, 57–69.

Frisén J, Lendahl U. (2001) Oh no, Notch again! *Bioessays,* 23, 307.

Fritsche J, Reber BF, Schindelholz B, Bandtlow CE. (1999) Differential cytoskeletal changes during growth cone collapse in response to hSema III and thrombin. *Mol. Cell. Neurosci.* 14, 398–418.

Frotscher M, Drakew A, Heimrich B. (2000) Role of afferent innervation and neuronal activity in dendritic development and spine maturation of fascia dentata granule cells. *Cereb. Cort.* 10, 946–951.

Fukuhara S, Mukai H, Munekata E. (1997) Activin A and all-trans-retinoic acid cooperatively enhanced the functional activity of L-type Ca^{2+} channels in the neuroblastoma C1300 cell line. *Biochem. Biophys. Res. Comm.* 241, 363–368.

Fukahori M, Ichimori K, Ishida H, Nakagawa H, Okino H. (1994) Nitric oxide reversibly suppresses xanthine oxidase activity. *Free Rad. Res.* 21, 203–212.

Gaidarov I, Keen JH. (1999) Phosphoinositide-AP-2 interactions required for targeting to plasma membrane clathrin-coated pits. *J. Cell Biol.* 146, 755–764.

Garey LJ, Ong WY. Patel TS, Kanani M, Davis A, Mortimer AM, Barnes TR, Hirsch SR. (1998) Reduced dendritic spine density on cerebral cortical pyramidal neurons in schizophrenia. *J. Neurol. Neurosurg. Psychiat.* 65, 446–453.

Garner CC, Nash J, Huganir RL. (2000) PDZ domains in synapse assembly and signalling. *TICB* 10, 274–280.

Geerlings A, López-Corcuera B, Aragón C. (2000) Characterization of the interactions between the glycine transporters GLYT1 and GLYT2 and the SNARE protein syntaxin 1A. *FEBS Lett.* 470, 51–54.

Gerlai R, McNamara A. (2000) Anesthesia induced retrograde amnesia is ameliorated by ephrin A5-IgG in mice: EphA receptor tyrosine kinases are involved in mammalian memory. *Behav. Br. Res.* 108, 133–143.

Gerst JL, Raina AK, Pirim I, McShea A, Harris PL, Siedlak SL, Takeda A, Petersen RB, Smith MA. (2000) Altered cell-matrix associated ADAM proteins in Alzheimer disease. *J. Neurosci. Res.* 59, 680–684.

Giannini AL, Vivanco MdM, Kypta RM (2000) Analysis of β-catenin aggregation and localization using GFP fusion proteins: nuclear import of α-catenin by the β-catenin/Tcf complex. *Exp. Cell Res.* 255, 207–220.

Gibson A, Futter, CE, Maxwell S, Allchin EH, Shipman M, Kraehenbuhl JP, Domingo D, Odorizzi G, Trowbridge IS, Hopkins CR. (1998) Sorting mechanisms regulating membrane protein traffic in the apical transcytotic pathway of polarized MDCK cells. *J. Cell Biol.* 143, 81–94.

Giger RJ, Cloutier JF, Sahay A, Prinjha RK, Levengood DV, Moore SE, Pickering S, Simmons D, Rastan S, Walsh FS, Kolodkin AL, Ginty DD, Gepert M. (2000) Neuropilin-2 is required in vivo for selective axon guidance responses to secreted semaporins. *Neuron* 25, 29–41.

Glantz LA, Lewis DA. (1995) Assessment of spine density on layer III pyramidal cells in the prefrontal cortex of schizophrenic patients. *Abst. Soc. Neurosci.* 21, 239.

Glantz LA, Lewis DA. (1997) Reduction of synaptophysin immunoreactivity in the pre-frontal cortex of subjects with schizophrenia. *Arch. Gen. Psychiat.* 54, 660–669 and 943–952.

Glantz LA, Lewis DA. (2000) Decreased dendritic spine density on prefrontal cortical pyramidal neurons in schizophrenia. *Arch. Gen. Psychiat.* 57, 65–73.

Glazner GW, Chan SL, Lu C, Mattson MP. (2000) Caspase-mediated degradation of AMPA receptor subunits: a mechanism for preventing excitotoxic necrosis and ensuring apoptosis. *J. Neurosci.* 20, 3641–3649.

Goldkorn T, Balaban N, Shannon M, Chea V, Matsukuma K, Gilchrist D, Wang H, Chan C. (1998) H_2O_2 acts on cellular membranes to generate ceramide signaling and initiate apoptosis in tracheobronchial epithelial cells. *J. Cell Sci.* 111, 3209–3220.

Goldman-Rakic PS, Selemon LD. (1997) Functional and anatomical aspects of prefrontal pathology in schizophrenia. *Schiz. Bull.* 23, 437–458.

Goodman MN, Reigh CW, Landreth GE. (1998) Physiological stress and nerve growth factor treatment regulate stress-activated protein kinase activity in PC12 cells. *J. Neurobiol.* 36, 537–549.

Gomperts SN, Carroll R. Malenka RC, Nicoll RA. (2000) Distinct roles for ionotropic and metabotropic glutamate receptors in the maturation of excitatory synapses. *J. Neurosci.* 20, 2229–2237.

Gottschalk W, Pozzo-Miller LD, Figurov A, Lu B. (1998) Presynaptic modulation of synaptic transmission and plasticity by brain-derived neurotrophic factor in the developing hippocampus. *J. Neurosci.* 18, 6830–6839.

Gould E, Cameron HA. (1997) Early NMDA receptor blockade impairs defensive behavior and increases cell proliferation in the dentate gyrus of developing rats. *Behav. Neurosci.* 111, 49–56.

Gould E, McEwan BS, Tanapat P, Galea LA, Fuchs E. (1997) Neurogenesis in the dentate gyrus of the adult tree shrew is regulated by psychosocial stress and NMDA receptor activation. *J. Neurosci.* 17, 2492–2498.

Grady EF, Gamp PD, Jones E, Baluk P, McDonald DM, Payan DG, Bunnett NW. (1996) Endocytosis and recycling of neurokinin 1 receptors in enteric neurons. *Neuroscience* 75, 1239–1254.

Greener T, Zhao X, Nojima H, Eisenberg E, Greene LE. (2000) Role of cyclin G-associated kinase in uncoating clathrin-coated vesicles from non-neuronal cells. *J. Biol. Chem.* 275, 1365–1370.

Greengard P, Allen PB, Nairn AC. (1999) Beyond the dopamine receptor: the DARPP-32/protein phosphatase-1 cascade. *Neuron* 23, 435–447.

Grieb TA, Burgess WH. (2000) The mitogenic activity of fibroblast growth factor-1 correlates with its internalization and limited proteolytic processing. *J. Cell. Physiol.* 184, 171–182.

Grimes ML. Zhou J, Beattie EC, Yuen EC, Hall DE, Valletta JS, Topp KS, LaVail JH, Bunnett NW, Mobley WC. (1996) Endocytosis of activated TrkA: evidence that nerve growth factor induces formation of signaling endosomes. *J. Neurosci.* 16, 7950–7964.

Grof S. (1963) Clinical and experimental study of central effects of adrenochrome. *J. Neuropsychiat.* 5, 33–50.

Grosse G, Grosse J, Tapp R, Kuchinke J, Gorsleben M, Fetter I, Höhne-Zell B, Gratzl M, Bergmann M. (1999) SNAP-25 requirement for dendritic growth of hippocampal neurons. *J. Neurosci. Res.* 56, 539–546.

Grover LM, Yan C. (1999) Evidence for involvement of group II/III metabotropic glutamate receptors in NMDA receptor-independent long-term potentiation in area CA1 of rat hippocampus. *J. Neurophysiol.* 82, 2956–2969.

Grumet M. (1997) Nr-CAM: a cell adhesion molecule with ligand and receptor functions. *Cell Tissue Res.* 290, 423–428.

Grune T, Davies KJ. (1997) Breakdown of oxidized proteins as a part of second antioxidant defenses in mammalian cells. *Biofactors* 6, 165–172.

Grünewald RA (1993) Ascorbic acid in the brain. *Br. Res. Rev.* 18, 123–133.

Gu F, Gruenberg J. (1999) Biogenesis of transport intermediates in the endocytic pathway. *FEBS Lett.* 452, 61–66.

Guidarelli A, Palomba L, Cantoni O. (2000) Peroxynitrite-mediated release of arachidonic acid from PC12 cells. *Brit. J. Pharmacol.* 129, 1539–1541.

Guidotti A, Pesold C, Costa E. (2000) New neurochemical markers for psychosis: a working hypothesis of their operation. *Neurochem. Res.* 25, 1207–1218.

Guo ZH, Mattson MP. (2000) Neurotrophic factors protect cortical synaptic terminals against amyloid- and oxidative stress-induced impairment of glucose transport, glutamate transport and mitochondrial function. *Cerebral Cortex* 10, 50–57.

Guzowski JF, Lyford GL, Stevenson GD, Houston FP, McGaugh JL, Worley PF, Barnes CA. (2000) Inhibition of activity-dependent arc protein expression in the rat hippocampus impairs the maintenance of long-term potentiation and the consolidation of long-term memory. *J. Neurosci.* 20, 3993–4001.

Hagler DJ Jr., Goda Y. (1998) Synaptic adhesion: the building blocks of memory? *Neuron* 20, 1059–1062.

Hahm S, Mizuno TM, Wu TJ, Wisor JP, Priest CA, Kozak CA, Boozer CN, Peng, B, McEvoy RC, Good P, Kelley KA, Takahashi JS, Pintar JE, Roberts JL, Mobbs CV, Salton SR. (1999) Targeted deletion of the Vgf gene indicates that the encoded secretory peptide precursor plays a novel role in the regulation of energy balance. *Neuron* 23, 537–548.

Hall AC, Lucas FR, Salinas PC. (2000) Axonal remodeling and synaptic differentiation in the cerebellum is regulated by WNT-7a signaling. *Cell* 100, 525–535.

Halpain S. (2000) Actin and the agile spine: how and when do dendritic spines dance? *TINS* 23, 141–146.

Halpain S, Hipolito A, Saffer L. (1998) Regulation of F-actin stability in dendritic spines by glutamate receptors and calcineurin. *J. Neurosci.* 18, 9835–9844.

Hammad SM, Barth JL, Knaak C, Argraves WS. (2000) Megalin acts in concert with cubilin to mediate endocytosis of high-density lipoproteins. *J. Biol. Chem.* 275, 12,003–12,008.

Han H, Yang S-H, Phillips WD. (2000) Overexpression of rapsyn modifies the intracellular trafficking of acetylcholine receptors. *J. Neurosci. Res.* 60, 155–163.

Hanson PI. (2000) Sec1 gets a grip on syntaxin. *Nature Struct. Biol.* 7, 347–349.

Harada S, Tachikawa H, Kawanishi Y. (2001) Glutathione S-transferase M1 gene deletion may be associated with susceptibility to certain forms of schizophrenia. *Biochem. Biophys. Res. Com.* 281, 267–271.

Harris KM. (1999) Structure, development, and plasticity of dendritic spines. *Curr. Opin. Neurobiol.* 9, 343–348.

Hassel B, Bråthe A. (2000) Neuronal pyruvate carboxylation supports formation of transmitter glutamate. *J. Neurosci.* 20, 1342–1347.

Haucke V. (2000) Dissecting the ins and outs of excitation; glutamate receptors on the move. *Nature Neuroscience* 3, 1230–1232.

He Y, Janssen WG, Rothstein JD, Morrison JH. (2000) Differential synaptic localization of the glutamate transporter EAAC1 and glutamate subunit GluR2 in the rat hippocampus. *J. Comp. Neurol.* 418, 255–269.

Hedberg TG, Stanton PK. (1996) Long-term plasticity in cingulate cortex requires both NMDA and metabotropic glutamate receptor activation. *Eur. J. Pharmacol.* 310, 19–27.

Heeren J. Weber W, Beisiegel U. (1999) Intracellular processing of endocytosed triglycer-ide-rich lipoproteins comprises both recycling and degradation. *J. Cell Sci.* 112, 349–359.

Henkel AW, Meiri H, Horstmann H, Lindau M, Almers W. (2000) Rhythmic opening and closing of vesicles during constitutive exo- and endocytosis in chromaffin cells. *Embo J.* 19, 84–93.

Henkel MK, Pott G, Henkel AW, Juliano L, Kam, C-M, Powers JC, Franzusoff A. (1999) Endocytic delivery of intramolecularly quenched substrates and inhibitors to the intracellular yeast Kex2 protease1. *Biochem. J.* 341, 445–452.

Herbert H, Saper CB. (1992) Organization of medullary adrenergic and noradrenergic projec-tions to the periaqueductal gray matter in the rat. *J. Comp. Neurol.* 315, 34–52.

Hevroni D, Rattner A, Bundman M, Lederfein D, Gabarah A, Mangelus M, Silverman MA, Kedar H, Naor C, Kornuc M, Hanoch T, Seger R, Theill LE, Nedivi E, Richter-Levin G, Citri Y. (1998) Hippocampal plasticity involves extensive gene induction and multiple cellular mechanisms. *J. Mol. Neurosci.* 10, 75–98.

Heynen AJ, Quinlan EM, Bae DC, Bear MF. (2000) Bidirectional, activity-dependent regula-tion of glutamate receptors in the adult hippocampus in vivo. *Neuron* 28, 527–536.

Hill E, van der Kaay J, Downes P, Smythe E. (2001) The role of dynamin and its binding partners in coated pit invagination and scission. *J. Cell Biol.* 152, 309–323.

Hipkiss AR. (1998) Carnosine, a protective, anti-ageing peptide? *Intern. J. Biochem. Cell. Biol.* 30, 863–868.

Hipkiss AR, Preston JE, Himsworth DTM, Worthington VC, Abbott NJ. (1997) Protective effects of carnosine against malondialdehyde-induced toxicity towards cultured rat brain endothelial cells. *Neurosci. Lett.* 238, 135–138.

Hirasawa A, Awaji T, Sugawara T, Tsujimoto A, Tsujimoto G. (1998) Differential mecha-nism for the cell surface sorting and agonist-promoted internalization of the alpha1B-adenoceptor. *Brit. J. Pharmacol.* 124, 55–62.

Hisanaga K, Sagar SM, Sharp FR. (1992) N-methyl-D-aspartate antagonists block fos-like protein expression induced via multiple signaling pathways in cultured cortical neurones. *J. Neurochem.* 58, 1836–1844.

Hoffer A, Osmond H, Smythies JR. (1994) Schizophrenia. A new approach. Part II. *J. Ment. Sci.* 100, 29–45.

Hoffman KB, Larson J, Bahr BA, Lynch G. (1998) Activation of NMDA receptors stimu-lates extracellular proteolysis of cell adhesion molecules in hippocampus. *Br. Res.* 811, 152–155.

Holscher C. (1997) Long-term potentiation: a good model for learning and memory? *Prog. Neuropsychopharmcol. Biol. Psychiat.* 21, 47–68.

Honer WG, Falkai P, Young C, Wang T, Xie J, Bonner J, Hu L, Boulianne GL, Luo Z, Trimble WS. (1997) Cingulate cortex synaptic terminal proteins and neural cell adhesion molecules in schizophrenia. *Neuroscience* 78, 99–110.

Hong K, Nishiyama M, Henley J, Tessier-Lavigne M, Poo M. (2000) Calcium signaling in the guidance of nerve growth by netrin-1. *Nature* 403, 93–98.

Höpker VH, Shewan D, Tessier-Lavigne M, Poo M, Holt C. (1999) Growth-cone attraction to netrin-1 is converted to repulsion by laminin-1. *Nature* 401, 69–73.

Horch HW, Krüttgen A, Portbury SD, Katz LC. (1999) Destabilization of cortical den-drites and spines by BDNF. *Neuron* 23, 353–364.

Horimoto N, Nabekura J, Ogawa T. (1997) Arachidonic acid activation of potassium channels in rat visual cortex neurons. *Neuroscience* 77, 661–671.

Horowitz A, Murakami M, Gao Y, Simons M. (1999) Phosphatidylinositol-4,5-biphos-phate mediates the interaction of syndecan-4 with protein kinase C. *Biochemistry* 38, 15,871–15,877.

Hortsch M. (2000) Structural and functional evolution of the L1 family; are four adhesion molecules better than one? *Mol. Cell. Neurosci.* 15, 1–10.

Horvitz JC. (2000) Mesolimbocortical and nigrostriatal dopamine responses to salient non-reward events. *Neuroscience* 96, 651–656.

Hoyt KR, Gallagher AJ, Hastings TG, Reynolds IJ. (1997) Characterization of hydrogen peroxide toxicity in cultured rat forebrain neurons. *Neurochem. Res.* 22, 333–340.

Hsueh YP, Wang TF, Yang FC, Sheng M. (2000) Nuclear translocation and transcription regulation by the membrane-associated guanylate kinase CASK/LIN-2. *Nature,* 404, 298–302.

Hu Y, Barzilai A, Chen, M, Bailey CH, Kandel ER. (1993) 5-HT and cAMP induce the formation of coated pits and vesicles and increase the expression of clathrin light chains in sensory neurons of aplysia. *Neuron* 10, 921–929.

Huang EP. (1999) Synaptic plasticity: regulated translation in dendrites. *Curr. Biol.* 9, R168–R170.

Huang Y-Y, Martin KC, Kandel ER. (2000) Both protein kinase A and mitogen-activated protein kinase are required in the amygdala for the macromolecular synthesis-dependent late phase of long-term potentiation. *J. neurosci.* 20, 6317–6325.

Hulet SW, Heyliger SO, Powers S, Connor JR. (2000) Oligodendrocyte progenitor cells internalize ferritin via clathrin-dependent receptor mediated endocytosis. *J. Neurosci. Res.* 61, 52–60.

Hülsmann S, Oku Y, Zhang W, Richter DW. (2000) Metabotropic glutamate receptors and blockade of glial Krebs cycle depress glycinergic synaptic currents of mouse hypoglossal motoneurons. *Eur. J. Neurosci.* 12, 239–266.

Hummel T, Krukkert K, Roos J, Davis G, Klämbt C. (2000) *Drosophila* Futsch/22C10 is a MAP1β-like protein required for dendritic and axonal development. *Neuron* 26, 357–370.

Humphries MJ. (2000) Integrin cell adhesion receptors and the concept of agonism. *TIPS* 21, 29–32.

Husi H, Ward MA, Choudhary JS, Blackstock WP, Grant SGN. (2000) Proteomic analysis of NMDA receptor-adhesion protein signaling complexes. *Nature Neuroscience* 3, 661–669.

Hussain NK, Yamabhai M, Ramjaun AR, Baranes D, O'Bryan JP, Der CJ, Kay BK, McPherson PS. (1999) Splice variants of intersectin are components of the endocytic machinery in neurons and nonneuronal cells. *J. Biol. Chem.* 274, 15,671–15,677.

Hutter D, Greene JJ. (2000) Influence of the cellular redox state on NF-κB-regulated gene expression. *J. Cell. Physiol.* 83, 45-52.

Hynes R (1999) Cell adhesion: old and new questions. *TICB* 12, M33-M37.

Iacovelli L, Sallese M, Mariggió S, de Blasi A. (1999) Regulation of G-protein-coupled receptor kinase subtypes by calcium sensor proteins. *FASEB J.* 13, 1–8.

Iacovitti L, Stull ND, Mishizen A. (1999) Neurotransmitters, KCl and antioxidants rescue striatal cells from apoptotic cell death in culture. *Brain Res,* 816, 276–285.

Ibata K, Fukuda M, Hamada T, Kabayama H, Mikoshiba K. (2000) Synaptotagmin 1V is present at the Golgi and distal parts of neurites. *J. Neurochem.* 74, 518–526.

Ignatova EG, Belcheva MM, Bohn LM, Neuman MC, Coscia CJ. (1999) Requirement of receptor internalization for opioid stimulation of mitogen-activated protein kinase; biochemical and immunofluorescence confocal microscopic evidence. *J. Neurosci.* 19, 56–63.

Iida M, Miyazaki I, Tanaka K, Kabuto H, Iwata-Ichikawa E, Ogawa N. (1999) Dopamine D2 receptor-mediated antioxidant and neuroprotective effects of ropinirole, a dopamine agonist. *Br. Res.* 838, 51–59.

Incardona JP, Eaton S. (2000) Cholesterol in signal transduction. *Curr. Opin. Cell Biol.* 12, 193–203.

Irie Y, Yamagata K, Gan Y, Miyamoto K, Do E, Kuo CH, Taira E, Miki N. (2000) Molecular cloning and characterization of Amida, a novel protein which interacts with a neuron-specific immediate early gene product arc, contains novel nuclear localization signals, and causes cell death in cultured cells. *J. Biol. Chem.* 275, 2647–2653.

Isaacson JS. (1999) Glutamate spillover mediates excitatory transmission in the rat olfactory bulb. *Neuron* 23, 377–384.

Ishihara I, Minami Y, Nishizaki T, Matsuoka T, Yamamura H. (2000) Activation of calpain precedes morphological alterations during hydrogen peroxide-induced apoptosis in neuronally differentiated mouse embryonal carcinoma P19 cell line. *Neurosci. Lett.* 279, 97–100.

Ishisaki A, Hayashi H, Suzuki S, Ozawa K, Mizukoshi E, Miyakawa K, Suzuki M, Imamura T. (2001) Glutathione S-transferase Pi is a dopamine-inducible suppressor of dopamine-induced apoptosis in PC12 cells. *J. Neurochem. 77, 1362–1371.*

Iwahori Y, Saito H, Torii K, Nishiyama N. (1997) Activin exerts a neurotrophic effect on cultured hippocampal neurons. *Brain Res.* 760, 52–58.

Jacobs B, Driscoll L, Schall M, (1997) Life-span dendritic and spine changes in areas 10 and 18 of human cortex: a quantitative Golgi study. *J. Comp. Neurol.* 386, 661–680.

Jakob U, Muse, W, Eser M, Bardwell JCA. (1999) Chaperone activity with a redox switch. *Cell* 96, 341–352.

Janáky R, Ogita K, Pasqualotto BA, Bains JS, Oja SS, Yoneda Y, Shaw CA. (1999) Glutathione and signal transduction in the mammalian CNS. *J. Neurochem.* 73, 889–902.

Jans DA, Hassan G. (1998) Nuclear targeting by growth factors, cytokines, and their receptors: a role in signaling? *Bioessays* 20, 400–411.

Janssen-Heininger YMW, Poynter ME, Baeuerle PA. (2000) Recent advances towards understanding redox mechanisms in the activation of nuclear factor κB. *Free Rad. Biol. Med.* 28, 1317–1327.

Jentsch S, Pyrowolakis G. (2000) Ubiquitin and its kin: how close are the family ties? *TCB* 10, 335–342.

Jess TJ, Belham CM, Thompson FJ, Scott PH, Plevin RJ, Gould GW. (1996) Phosphatidylinositol 3′-kinase, but not p70 ribosmal S6 kinase, is involved in membrane protein recycling: wortmannin inhibits glucose transport and downregulates cell-surface transferrin receptor numbers independently of any effect on fluid-phase endocytosis in fibroblasts. *Cell. Signal.* 8, 297–304.

Ji RR, Böse CM, Lesuisse C, Qiu D, Huang JC, Zhang Q, Rupp F. (1998) Specific agrin isoforms induce cAMP response binding protein phosphorylation in hippocampal neurons. *J. Neurosci.* 18, 9695–9702.

Jockers R, Angers S, Da Silva A, Benaroch P, Strosberg AD, Bouvier M, Marullo S. (1999) Beta(2)-adrenergic receptor down-regulation. Evidence for a pathway that does not require endocytosis. *J. Biol. Chem.* 274, 28,900–28,908.

Jovanovic JN, Czernik AJ, Fienberg AA, Greengard P, Sihra TS. (2000) Synapsins as mediators of BDNF-enhanced neurotransmitter release. *Nature Neuroscience* 3, 323–329.

Jugloff DG, Khanna R, Schlichter LC, Jones OT. (2000) Internalization of the Kv1.4 potassium channel is suppressed by clustering interactions with PSD-95. *J. Biol. Chem.* 275, 1357–1364.

Juraska JM, Greenough WT, Elliott C, Mack KJ, Berkowitz R. (1980) Plasticity in adult rat visual cortex: an examination of several cell populations after differential rearing. *Behav. Neural Biol.* 29, 157–167.

Kalus P, Müller TJ, Zuschratter W, Senitz D. (2000) The dendritic architecture of prefrontal pyramidal neurons in schizophrenic patients. *NeuroReport* 11, 3621–3625.

Kamata H, Hirata H. (1999) Redox regulation of cellular signaling. *Cell. Signal.* 11, 1–14.

Kamata H, Shibukawa Y, Oka SI, Hirata H. (2000) Epidermal growth factor receptor is modulated by redox through multiple mechanisms. Effects of reductants and H_2O_2. *Eur. J. Biochem.* 267, 1933–1944.

Kamiguchi H, Lemmon V. (2000) Recycling of the cell adhesion molecule L1 in axonal growth cones. *J. Neurosci.* 20, 3676–3686.

Kammermeier PJ, Xiao B, Tu JC, Worley PF, Ikeda SR. (2000) Homer protein regulates coupling of group I metabotropic glutamate receptors to N-type calcium and M-type potassium channels. *J. Neurosci.* 20, 7238–7245.

Karanth S, Yu WH, Walczewska A, Mastronardi C, McCann SM. (2000) Ascorbic acid acts as an inhibitory transmitter in the hypothalamus to inhibit stimulated luteinizing hormone-releasing hormone release by scavenging nitric oxide. *PNAS USA* 97, 1891–1896.

Karson CN, Griffin WS, Mrak RE, Sturner WQ, Shillcutt S, Guggenheim FG. (1997) Reduced levels of synaptic proteins in the prefrontal cortex in schizophrenia, *Biol. Psychiatr.* 41, 59S.

Kato H, Sugino N, Takiguchi S, Kashida S, Nakamura Y. (1997) Roles of reactive oxygen species in regulation of luteal function. *Rev. Reproduct.* 2, 81–83.

Katsuki H, Noguchi K, Matsuki N. (1998) Modulation of the synaptic plasticity by hydrogen peroxide in the hippocampus. *Abst. Soc. Neurosci.* 24, 1070.

Kaufmann WE, Worley PF, Pegg J, Bremer M, Isakson P. (1996) Cox-2, a synaptically induced enzyme, is expressed by excitatory neurons at postsynaptic sites in rat cerebral cortex. PNAS USA 93, 2317–2321.

Kawakami T, Suzuki T, Baek SH, Chung CH, Kawasaki H, Hirano H, Ichiyama A, Omata M, Tanaka K. (1999) Isolation and characterization of cytosolic and membrane-bound deubiquitinylating enzymes from bovine brain. *J. Biochem.* 126, 612–623.

Keller A, Arissian K, Asanuma H. (1992) Synaptic proliferation in the motor cortex of adult cats after long-term thalamic stimulation. *J. Neurophysiol.* 68, 295–308.

Keller JN, Hanni KB, Markesbery WR. (2000) Impaired proteosome function in Alzheimer's disease. *J. Neurochem.* 75, 436–439.

Kelly RB. (1999) Deconstructing membrane traffic. *TICB* 9, M29-M32.

Kietzmann T and Fandrey J. (2000) Oxygen radicals as messengers in oxygen-dependent gene expression. *NIPS* 15, 202–208.

Kikuchi A. (2000) Regulation of β-catenin in the Wnt pathway. *Biochem. Biophys. Res. Com.* 268, 243–248.

Kimura K, Hattori S, Kabuyama Y, Shizawa Y, Takayanagi J, Nakamura S, Toki S, Matsuda Y, Onodera K, Fukui Y. (1994) Neurite outgrowth of PC12 cells is suppressed by wortmannin, a specific inhibitor of phosphatidylinositol 3-kinase. *J. Biol. Chem.* 269. 18,961–18,967.

Kitayama I, Yaga T, Kayahara T, Nakano K, Murase S, Otani M, Nomura J. (1997) Long-term stress degenerates, but imipramine regenerates, noradrenergic axons in the rat cerebral cortex. *Biol. Psychiat.* 42, 687–696.

Kiyatkin EA, Rebec GV. (1998) Ascorbate modulates glutamate-induced excitations of striatal neurons. *Brain Res.* 812, 14–22.

Kjome JR, Swenson KA, Johnson MN, Bordayo EZ, Anderson LE, Klevan LC, Fraticelli AI, Aldrich SL, Fawcett JR, Venters HD Jr., Ala TA, Frey WH 2nd. (1998) Inhibition of antagonist and agonist binding to the human brain muscarinic receptor by arachidonic acid. *J. Mol. Neurosci.* 10, 209–217.

Kleim JA, Lussnig E, Schwartz ER, Comery TA, Greenough WT. (1996) Synaptogenesis and Fos expression in the motor cortex of the adult rat after motor skill learning. *J. Neurosci.* 16, 4529–4535.

Kleim JA, Vij K, Ballard DH, Greenough WT. (1997) Learning-dependent synaptic modifications in the cerebellar cortex of the adult rat persist for at least four weeks. *J. Neurosci.* 17, 717–721.

Ko GY, Kelly PT. (1999) Nitric oxide acts as a postsynaptic signaling molecule in calcium/calmodulin-induced synaptic potentiation in hippocampal CA1 pyramidal neurons. *J. Neurosci.* 19, 6784–6794.

Ko KS, McCulloch CAG. (2000) Partners in protection: interdependence of cytoskeleton and plasma membrane in adaptions to applied forces. *J. Membr. Biol.* 174, 85–95.

Kobayashi T, Gu F, Gruenberg J. (1998) Lipids, lipid domains, and lipid-protein interactions in endocytic membrane traffic. *Sem. Cell. Develop. Biol.* 9, 517–526.

Koenig JA, Edwardson JM. (1997) Endocytosis and recycling of G-protein-coupled receptors. *TIPS* 18, 276–287.

Kokubo H, Lemere CA, Yamaguchi H. (2000) Localization of flotillins in human brain and their accumulation with the progression of Alzheimer's disease. *Neurosci. Lett.* 290, 93–96.

Kolesnick RN, Goñi FM, Alonso A. (2000) Compartmentalization of ceramide signaling: physical foundations and biological effects. *J. Cell. Physiol.* 184, 285–300.

Kolkova K, Novitskaya V, Pedersen N, Berezin V, Bock E. (2000) Neural cell adhesion molecule-stimulated neurite outgrowth depends on activation of protein kinase C and the Ras-mitogen-activated protein kinase pathway. *J. Neurosci.* 20, 2238–2246.

Koppenol WH. (1998) The basic chemistry of nitrogen monoxide and peroxynitrite. *Free Rad. Biol. Med.* 25, 385–391.

Kordeli E. (2000) The spectrin-based skeleton at the postsynaptic membrane of the neuromuscular junction. *Micros. Res. Tech.* 49, 101–107.

Korkotian E, Segal M. (1999) Bidirectional regulation of dendritic spine dimensions by glutamate receptors. *NeuroReport* 10, 2875–2877.

Kornitzer D, Ciechanover A. (2000) Modes of regulation of ubiquitin-mediated degradation. *J. Cell. Physiol.* 182, 1–11.

Kossel AH, Williams CV, Schweitzer M, Kater SB. (1997) Afferent innervation influences the development of dendritic branches and spines via both activity-dependent and non-activity-dependent mechanisms. *J. Neurosci.* 17, 6314–6324.

Kötter R. (1994) Postsynaptic integration of glutamatergic and dopaminergic signals in the striatum. *Prog. Neurobiol.* 44, 163–196.

Koulen P. (1999) Clustering of neurotransmitter receptors in the mammalian retina. *J. Memb. Biol.* 171, 97–105.

Kozlov MM. (1999) Dynamin: possible mechanism of "Pinchase" action. *Biophys. J.* 77, 604–616.

Krugmann S, Welch H. (1998) P13-kinase. *Curr. Biol.* 8, R828.

Kullmann DM, Asztely F. (1998) Extrasynaptic glutamate spillover in the hippocampus: evidence and implications. *TINS* 21, 8–14.

Kunizuka H, Kinouchi H, Arai K, Izaki K, Mikawa S, Kamii H, Sugawara T, Suzuki A, Mizoi K, Yoshimoto T. (1999) Activation of Arc gene, a dendritic immediate early gene, by middle cerebral artery occlusion in rat brain. *NeuroReport* 10, 1717–1722.

Kunz S, Spirig M, Ginsburg C, Buchstaller A, Berger P, Lanz R, Rader C, Vogt L, Kunz B, Sonderegger P. (1998) Neurite fasciculation mediated by complexes of axonin-1 and Ng cell adhesion molecule. *J. Cell Biol.* 143, 1673–1690.

Kurashima K, Szabó EZ, Lukacs G, Orlowski J, Grinstein S. (1998) Endosomal recycling of the Na$^+$/H$^+$ exchanger NHE3 isoform is regulated by the phosphatidylinositol 3-kinase pathway. *J. Biol. Chem.* 273, 20828–20836.

Kwon JH, Eves EM, Farrell S, Segovia J, Tobin AJ, Wainer BH, Downen M. (1996) beta-adrenergic receptor activation promotes process outgrowth in an embryonic rat basal forebrain cell line and in primary neurons. *Eur. J. Neurosci.* 8, 2042–2055.

Lah JJ, Levey AI. (2000) Endogenous presenilin-1 targets to endocytic rather than biosynthetic compartments. *Mol. Cell Neurosci.* 16, 111–126.

Lam HHD, Bhardwaj A, O'Connell MT, Hanley DF, Traystman RJ, Sofroniew MV. (1998) Nerve growth factor rapidly suppresses basal, NMDA-evoked, and AMPA-evoked nitric oxide synthase activity in rat hippocampus in vivo. *PNAS USA* 95, 10,926–10,931.

Laming PR, Kimelberg H, Robinson S, Salm A, Hawrylak N, Müller C, Roots B, Ng K. (2000) Neuron-glial interactions and behaviour. *Neurosci. Biobehav. Rev.* 24, 295–340.

Lancelot E, Callebert J, Lerouet D, Revaud M-L, Boulu RG, Plotkine M. (1995) Role of the L-arginine-nitric oxide pathway in the basal hydroxyl radical production in the striatum by awake rats as measured by brain microdialysis. *Neurosci. Lett.* 202, 21–24.

Landén M, Davidsson P, Gottfries C-G, Grenfeldt B, Stridsberg M, Blennow K. (1999) Reduction of the small synaptic vesicle protein synaptophysin but not the large dense core chromogranins in the left thalamus of subjects with schizophrenia. *Biol. Psychiatr.* 46, 1698–1702.

Lander HM. (1997) An essential role for free radicals and derived species in signal transduction. *FASEB J.* 11, 118–124.

Langeveld CH, Schepens E, Stoof JC, Bast A, Drukarch B. (1995) Differential sensitivity to hydrogen peroxide of dopaminergic and noradrenergic neurotransmission in rat brain slices. *Free Rad. Biol. Med.* 19, 209–217.

Laporte SA, Oakley RH, Zhang J, Holt JA, Ferguson SSG, Caron MG, Barak LS. (1999) The β2-adrenergic receptor/β-arrestin complex recruits the clathrin adaptor AP-2 during endocytosis. *PNAS USA* 96, 3712–3717.

Larsen CN, Krantz BA. Wilkinson KD. (1998) Substrate specificity of deubiquinating enzymes: ubiquitin C-terminal hydrolases. *Biochemistry* 37, 3358–3368.

Laughlan SB. (1999) Dendritic integration makes sense of the world. *Curr. Biol.* 9, R15–R17.

Laux T, Fukami K, Thelen M, Golub T, Frey D, Caroni P. (2000) GAP43, MARCKS, and CAP23 modulate PI(4,5)P(2) at plasmolemmal rafts, and regulate cell cortex actin dynamics through a common mechanism. J. Cell Biol. 149, 1455–1472.

Le TL, Yap AS, Stow JL. (1999) Recycling of E-cadherin: a potential mechanism for regulating cadherin dynamics. *J. Cell Biol.* 146, 219–232.

Le PU, Benlimame N, Lagana A, Raz A, Nabi IR. (2000) Clathron-mediated endocytosis and recycling of autocrine motility factor receptor to fibronectin fibrils is a limiting factor for NIH-3T3 cell motility. *J. Cell Sci.* 113, 3227–3240.

Lee Z-W, Kweon S-M, Kim S-J, Kim J-H, Cheong C, Park Y-M, Ha K-S. (2000) The essential role of H_2O_2 in the regulation of intracellular Ca^{2+} by epidermal growth factor in rat-2 fibroblasts. *Cell. Sig.* 12, 91–98.

Lei SZ, Pan Z-H, Aggarwal K, Chen HS, Hartman J, Sucher NJ, Lipton SA. (1992) Effect of nitric oxide production on the redox modulatory site of the NMDA receptor-channel complex. *Neuron* 8, 1087–1099.

Lein ES, Shatz CJ. (2000) Rapid regulation of brain-derived neurotrophic factor mRNA within eye-specific circuits during ocular dominance column formation *J. Neurosci.* 20, 1470–1483.

Lemmon SK, Traub LM. (2000) Sorting in the endosomal system in yeast and animal cells. *Curr. Opin. Cell Biol.* 12, 457–466.

Leof EB. (2000) Growth factor receptor signalling: location, location, location. *TICB* 10, 343–348.

Leveque C, Boudier JA, Takahashi M, Seagar M. (2000a) Calcium-dependent dissociation of synaptotagmin from synaptic SNARE complexes. *J. Neurochem.* 74, 367–374.

Leveque J-C, Macías W, Rajadyakska A, Carlson RR, Barczak A, Kang S, Li X-M, Coyle JT, Huganir RL, Heckers S, Konradi C. (2000b) Intracellular modulation of NMDA receptor function by antipsychotic drugs. *J. Neurosci.* 20, 4011–4020.

Levine MS, Cepeda C. (1998) Dopamine modulation of responses mediated by excitatory amino acids in the neostriatum. *Adv. Pharmacol.* 42, 723–729.

Levine ES, Crozier RA, Black IB, Plummer MR. (1998) Brain-derived neurotrophic factor modulates hippocampal synaptic transmission by increasing N-methyl-D-aspartic acid receptor activity. *PNAS USA* 95, 10235–10239.

Levkowitz G, Waterman H, Ettenberg SA, Katz M, Tsygankov A, Alroy I, Lavi S, Iwai K, Reiss Y, Ciechanover A, Lipkowitz S, Yarden Y. (1999) Ubiquitin ligase activity and tyrosine phosphorylation underlie suppression of growth factor signaling to c-Cbl/Sli-1. *Mol. Cell.* 4, 1029–1040.

Lew JY, Matusumoto Y, Pearson J, Goldstein M, Hökfelt T, Fuxe K. (1977) Localization and characterization of phenylethanolamine N-methyl transferase in the brain of various mammalian species. *Brain Res.* 119, 199–210.

Lewis P. Lentz TL. (1999) Rabies virus entry into cultured rat hippocampal neurons. *J. Neurocytol.* 27, 559–573.

Li A-J, Katafuchi T, Oda S, Hori T. Oomura Y. (1997) Interleukin-6 inhibits long-term potentiation in rat hippocampal slices. *Brain Res.* 748, 30–38.

Lidow MS, Song Z-M, Castner SA, Allen PB, Greengard P, Goldman-Rakic P. S. (2001) Antipsychotic treatment induces alterations in dendrite- and spine-associated proteins in dopamine-rich areas of the primate cerebral cortex. *Biol. Psychiat.* 49, 1–12.

Lieb K, Andrae J, Reisert I, Pilgrim C. (1995) Neurotoxicity of dopamine and protective effects of the NMDA receptor antagonist AP-5 differ between male and female neurons. *Exp. Neurol.* 134, 222–229.

Lieberman DN, Mody I. (1994) Regulation of NMDA channel function by endogenous Ca^{2+}-dependent phosphatase. *Nature* 369, 235–239.

Lin JW, Ju W, Foster K, Lee S, Ahmadian G, Wyszynski M, Wang YT, Sheng M. (2000) Distinct molecular mechanisms and divergent endocytotic pathways of AMPA receptor internalization. *Nature Neuroscience* 3, 1282–1290.

Lindén A-M, Väisänen J, Lakso M, Nawa H, Wong G, Castrén E. (2000) Expression of neurotrophins BDNF and NT-3, and their receptors in rat brain after administration of antipsychotic and psychotrophic agents. *J. Mol. Neurosci.* 14, 27–37.

Linden DJ. (1999) The return of the spike: postsynaptic action potentials and the induction of LTP and LTD. *Neuron* 22, 661–666.

Liochev SI and Fridovich I. (1999) Superoxide and iron: partners in crime. *Iubmb Life* 48, 157–161.

Lipinski P, Drapier JC, Oliviera L, Retmanska H, Sochanowicz B, Kruszewski M. (2000) Intracellular iron status as a hallmark of mammalian cell susceptibility to oxidative stress: a study of L5178Y mouse lymphoma cell lines differentially sensitive to H_2O_2. *Blood*, 95, 2960–2966.

Lissin DV, Carroll RC, Nicoll RA, Malenka RC, von Zastrow M. (1999) Rapid, activation-induced redistribution of ionotropic glutamate receptors in cultured hippocampal neurons. *J. Neurosci.* 19, 1263–1272.

Lledo P-M, Zhang X, Südhof TC, Malenka RC, Nicoll RA. (1999) Postsynaptic membrane fusion and long-term potentiation. *Science* 279, 399–403.

Lippard SJ. (1999) Free copper ions in the cell? *Science* 284, 748–749.

Liu J, Mori A. (1993) Monoamine metabolism provides an antioxidant defense in the brain against oxidant- and free radical-induced damage. *Exp. Neurol.* 134, 222–229.

Lodovichi C, Berardi N, Pizzorusso T, Maffei L. (2000) Effects of neurotrophins on cortical plasticity: same or different? *J. Neurosci.* 20, 2155–2165.

Lozovaya NA, Kopanitsa MV, Boychuk YA, Krishtal OA. (1999) Enhancement of glutamate release uncovers spillover-mediated transmission by N-methyl-D-aspartate receptors in the rat hippocampus. *Neuroscience* 91, 1321–1330.

Lu Y-F, Kandel ER, Hawkins RD. (1999) Nitric oxide signaling contributes to late-phase LTP and CREB phosphorylation in the hippocampus. *J. Neurosci* 19, 10250–10261.

Lu W-Y, Jackson MF, Bai D, Orser BA, MacDonald JF. (2000) In CA1 pyramidal neurons of the hippocampus protein kinase C regulates calcium-dependent inactivation of NMDA receptors. *J. Neurosci.* 20, 4452–4461.

Luo Y, Hattori A, Munoz J, Qin Z-H, Roth GS. (1999) Intrastriatal dopamine injection induces apoptosis through oxidation-involved activation of transciption factors AP-1 and NF-κB in rats. *Mol. Pharmacol.* 56, 254–264.

Lüscher C, Nicoll RA, Malenka RC, Muller D. (2000) Synaptic plasticity and dynamic modulation of the post-synaptic membrane. *Nature Neuroscience* 3, 545–550.

Lustig M, Sakurai T, Grumet M. (1999) Nr-CAM promotes neurite outgrowth from peripheral ganglia by a mechanism involving axonin-1 as a neuronal receptor. *Develop. Biol.* 209, 340–351.

Luzio JP, Rous BA, Bright NA, Pryor PR, Mullock BM, Piper RC. (2000) Lysosome-endosome fusion and lysosome biogenesis. *J. Cell Sci.* 113, 1515–1524.

Lysakowski A, Figueras H, Price SD, Peng Y-Y. (1999) Dense-cored vesicles, smooth endoplasmic reticulum, and mitochondria are closely associated with non-specialized parts of plasma membrane of nerve terminals: implications for exocytosis and calcium buffering by intraterminal organelles. *J. Comp. Neurol.* 403, 378–390.

Ma L, Zablow L, Kandel ER, Siegelbaum SA. (1999) Cyclic AMP induces functional presynaptic boutons in hippocampal CA3-CA1 neuronal cultures. *Nature. Neuroscience* 2, 24–30.

MacConell LA, Widger AE, Barth-Hall S, Roberts VJ. (1998) Expression of activin and follistatin in the rat hypothalamus: anatomical association with gonadotropin-releasing hormone neurons and possible role of central activin in the regulation of luteinizing hormone release. *Endocrine* 9, 233–241.

Magistretti PJ, Pellerin L. (1999) Astrocytes couple synaptic activity to glucose utilization in the brain. *NIPS* 14, 177–181.

Mailly F, Marin P, Israël M, Glowinski J, Prémont J. (1999) Increase in external glutamate and NMDA receptor activation contribute to H2O2-induced neuronal apoptosis. *J. Neurochem.* 73, 1181–1188.

Maletic-Savatic M, Malinow R, Svoboda K. (1999) Rapid dendritic morphogenesis in CA1 hippocampal dendrites induced by synaptic activity. *Science* 283, 1923–1927.

Man H-Y, Lin JW, Ju WH, Ahmadian G, Liu L, Becker LE, Sheng M, Wang YT. (2000) Regulation of AMPA receptor-mediated synaptic transmission by clathrin-dependent receptor internalization. *Neuron* 25, 649–662.

Manabe T, Aiba A, Yamada A, Ichise I, Sakagami H, Kondo H, Katsuki M. (2000) Regulation of long-term potentiation by H-ras through NMDA receptor phosphorylation. *J. Neurosci.* 20, 2504–2511.

Mangoura D. (1997) Mu-opioids activate tyrosine kinase focal adhesion kinase and regulate cortical cytoskeleton proteins cortactin and vinculin in chick embryonic neurons. *J. Neurosci. Res.* 50, 391–401.

Manna SK, Aggarwal BB. (2000) Wortmannin inhibits activation of nuclear transcription factors NF-κB and activated protein1 induced by lipopolysaccharide and phorbol ester. *FEBS Lett.* 473, 113–118.

Martin SJ, Morris RGM. (2001) Cortical plasticity: Its all the range. *Curr. Biol.* 11, R57–R59

Maruyama W, Dostert P, Naoi M. (1995) Dopamine-derived 1methyl-6,7-dihydroisoquinolines as hydroxyl radical promoters and scavengers in the rat brain; in vivo and in vitro studies. *J. Neurochem.* 64, 2635–2643.

Masino SA, Mesches MH, Bickford PC, Dunwiddie TV. (1999) Acute peroxide treatment of rat hippocampal slices induces adenosine-mediated inhibition of excitatory transmission in area CA1. *Neurosci. Lett.* 274, 91–94.

Matsumoto K, Yobimoto K, Huong NTT, Abdel-Fattah M, Van Hien T, Watanabe H. (1999) Psychological stress-induced enhancement of brain lipid peroxidation via nitric oxide systems and its modulation by anxiolytic and anxiogenic drugs in mice. *Brain Res.* 839, 74–84.

Mattson MP, Goodman Y, Luo H, Fu W, Furukawa K. (1997) Activation of NF-κB protects hippocampal neurons against oxidative stress-induced apoptosis: evidence for induction of manganese superoxide dismutase and suppression of peroxynitrite production and protein tyrosinase nitration. *J. Neurosci. Res.* 49, 681–697.

Matus A. (1999) Postsynaptic actin and neuronal plasticity. *Curr. Opin. Neurobiol.* 9, 561–565.

Maudsley S, Pierce KL, Zamah AM, Miller WE, Ahn S, Daaka Y, Lefkowitz RJ, Luttrell LM. (2000) The beta(2)-adrenergic receptor mediates extracellular signal-regulated kinase activation via assembly of a multi-receptor complex with the epidermal growth factor receptor. *J. Biol. Chem.* 275, 9572–9580.

Maximov A, Südhof TC, Bezprozvanny I. (1999) Association of neuronal calcium channels with modular adaptor proteins. *J. Biol. Chem.* 274, 24,453–24,456.

McAllister AK, Katz LC, Lo DC. (1996) Neurotrophin regulation of cortical dendritic growth requires activity. *Neuron* 17, 1057–1064.

McConalogue K, Grady EF, Minnis J, Balestra B, Tonini M, Brecha NC, Bunnett NW, Sternini C. (1999) Activation and internalization of the μ-opoid receptor by the newly discovered endogenous agonists, endomorphin-1 and endomorphin-2. *Neuroscience* 90, 1051–1059.

McDonald PH, Cote NL. Lin FT, Premont RT, Pitcher JA, Lefkowitz RJ. (1999) Identification of NSF as a beta-arrestin1-binding protein. Implications for beta2-adrenergic regulation. *J. Biol. Chem.* 274, 10677–10680.

McKinney RA, Capogna M, Dürr R, Gähwiler BH, Thompson SM. (1999) Minature synaptic events maintain dendritic spines via AMPA receptor activation. *Nature. Neuroscience.* 2, 44–49.

McLachlan DRC, Dalton AJ, Kruck TPA, Bell MY, Smith WL, Kalow W, Andrews DF. (1991) Intramuscular desferrioxamine in patients with Alzheimer's disease. *Lancet* 337, 1304–1308.

McLean Bolton M, Pittman AJ, Lo DC. (2000) Brain-derived neurotrophic factor differentially regulates excitatory and inhibitory synaptic transmission in hippocampal cultures. *J. Neurosci.* 20, 3221–3232.

McNiven MA, Cao H, Pitts KR, Yoon Y. (2000) The dynamin family of mechanoenzymes: pinching in new places. *TIBS* 25, 115–120.

Mednikova YS, Karnup SV, Loseva EV. (1998) Cholinergic excitation of dendrites in neocortical neurons. *Neuroscience* 87, 783–796.

Menegon A, Burgaya F, Baudot P, Dunlap DD, Girault JA, Valtorta F. (1999) FAK⁺ and PYK2/CAKbeta, two related tyrosine kinases highly expressed in the central nervous system: similarities and differences in the expression pattern. *Eur. J. Neurosci.* 11, 3777–3788.

Meyer T, Shen K. (2000) In and out of the postsynaptic region: signalling proteins on the move. *TICB* 10, 238–244.

Michel PP, Hefti F. (1990) Toxicity of 6-hydroxy dopamine and dopamine for dopaminergic neurons in culture. *J. Neurosci. Res.* 26, 428–435.

Micheva KD, Kay BK, McPherson PS. (1997) Synaptojanin forms two separate complexes in the nerve terminal. Interactions with endophilin and amphysin. *J. Biol. Chem.* 272, 27,239–27,245.

Milby KH, Mefford IN, Chey W, Adams RN. (1981) In vitro and in vivo depolarization coupled efflux of ascorbic acid in rat brain preparations. *Brain. Res. Bull.* 7, 237–242.

Mills LR, Morris CE. (1998) Neuronal plasma membrane dynamics evoked by osmomechanical perturbations. *J. Membr. Biol.* 166, 223–251

Milzani A, Dalledonne I. (1999) Effects of chlorpromazine on actin polymerization; slackening of filament elongation and filament annealing. *Arch. Biochem. Biophys.* 369, 59–67.

Ming G, Song H, Beringer B, Inagaki N, Tessier-Lavigne M, Poo M. (1999) Phospholipase C-gamma and phosphoinositide 3-kinase mediate cytoplasmic signaling in nerve growth cone guidance. *Neuron* 23, 139–148.

Mirnics K, Middleton FA, Marquez A, Lewis DA, Levitt P. (2000) Molecular characterization of schizophrenia reviewed by microassay analysis of gene expression in prefrontal cortex. *Neuron* 28, 53–67.

Missler M, Fernandez-Chacon R, Südhof TC. (1998) The making of neurexins. *J. Neurochem.* 71, 1339–1347.

Mizusawa H, Ishii T, Bannai S. (2000) Peroxiredoxin I (macrophage 23 kDa stress protein) is highly and widely expressed in the rat nervous system. *Neurosci. Lett.* 283, 57–60.

Mochida S. (2000) Protein-protein interactions in neurotransmitter release. *Neurosci. Res.* 36, 175–182.

Molina-Holgado F, Lledó A, Guaza C. (1995) Evidence for cyclooxygenase activation by nitric oxide in astrocytes. *Glia* 15, 167–172.

Monje ML. Chatten-Brown J, Hye SE, Raley-Susman KM. (2000) Free radicals are involved in the damage to protein synthesis after anoxia/aglycemia and NMDA exposure. *Brain Res.* 857, 172–182.

Morales M, Goda Y. (1999) Nomadic AMPA receptors and LTP. *Neuron* 23, 431–434.

Morinaka K, Koyama S, Nakashima S, Hinoi T, Okawa K, Iwamatsu A, Kikuchi A. (1999) Epsin binds to the EH domain of POB1 and regulates receptor-mediated endocytosis. *Oncogene* 18, 5915–5922.

Mufson EJ, Ma SY, Cochran EJ, Bennett DA, Beckett LA, Jaffar S, Savagovi HU, Kordower JH. (2000) Loss of nucleus basalis neurons containing trkA immunoreactivity in individuals with mild cognitive impairment and early Alzheimer's disease. *J. Comp. Neurol.* 427, 19–30.

Mukherjee S, Ghosh RN, Maxfield FR. (1997) Endocytosis. *Physiol. Rev.* 77, 759–803.

Mukhin YV, Garnovskya MN, Collingsworth G, Grewal JS, Pendergrass D, Nagai T, Pinckney S, Greene EL, Raymond JR. (2000) 5-hydroxytryptamine1A receptor/Giβγ stimulates mitogen-activated protein kinase via NAD(P)H oxidase and reactive oxygen species upstream of src in Chinese hamster ovary fibroblasts. *Biochem. J.* 347, 61–67.

Mundell SJ, Kelly E. (1998) The effect of inhibitors of receptor internalization on the desensitization and resensitization of three Gs-coupled receptor responses. *Brit. J. Pharmacol.* 125, 1594–1600.

Murase S, Schuman EM. (1999) The role of cell adhesion molecules in synaptic plasticity and memory. *Curr. Opin. Cell Biol.* 11, 549–553.

Murphy DD, Cole NB, Greenberger V, Segal M. (1998a) Estradiol increases dendritic spine density by reducing GABA neurotransmission in hippocampal neurons. *J. Neurosci.* 18, 2550–2559.

Murphy DD, Cole NB, Segal M. (1998b) Brain-derived neurotrophic factor mediates estradiol-induced dendritic spine formation in hippocampal neurons. *PNAS USA* 95, 11,412–11,417.

Murthy VN. (1999) Getting the membrane into shape for endocytosis. *Neuron* 24, 2–4.

Nagatsu I, Ikemoto K, Takeuchi T, Arai R, Karasawa N, Fujii T, Nagatsu T. (1998) Phenyl-ethonalamine-*N*-methyl transferase-immunoreactive nerve terminals afferent to the mouse substantia nigra. *Neurosci. Lett.* 245, 41–44.

Naisbitt S, Kim E, Tu JC, Xiao B, Sala C, Valtschanoff J, Weinberg RJ, Worley PF, Sheng M. (1999) Shank, a novel family of postsynaptic density proteins that binds to the NMDA receptor/PSD-95/GKAP complex and cortactin. *Neuron* 23, 569–582.

Nakashima S, Morinaka K, Koyama S, Ikeda M, Kishida M, Okawa K, Iwamatsu A, Kishida S, Kikuchi A. (1999) Small G protein Ral and its downstream molecules regulate endocytosis of EGF and insulin receptors. *Embo J.* 18, 3629–3642.

Narisawa-Saito M, Carnahan J, Araki K, Yamaguchi T, Nawa H. (1999) Brain-derived neurotrophic factor regulates the expression of AMPA receptor proteins in neocortical neurons. *Neuroscience* 88, 1009–1014.

Neve RL, Coopersmith R, McPhie DL, Santeufemio C, Pratt KG, Murphy CJ, Lynn SD. (1998) The neuronal growth-associated protein GAP-43 interacts with rabaptin-5 and participates in endocytosis. *J. Neurosci.* 18, 7757–7767.

Niblock MM, Brunso-Bechtold JK, Riddle DR. (2000) Insulin-like growth factor I stimulates dendritic growth in primary sensory cortex. *J. Neurosci.* 20, 4165–4176.

Nichol JA, Hutter OF. (1996) Tensile strength and dilatational elasticity of giant sarcolemmal vesicles shed from rabbit muscle. *J. Physiol.* 493, 187–198.

Nirenberg MJ, Chan J, Vaughan RA, Uhl GR, Kuhar MJ, Pickel VM. (1997) Immunogold localization of the dopamine transporter; an ultrastructural study of the rat ventral tegmental area. *J. Neurosci.* 17, 5255–5262.

Nishizaki T, Nomura T, Matsuoka T. Tsujishita Y. (1999) Arachidonic acid as a messenger for the expression of long-term potentiation. *Biochem. Biophys. Res. Comm.* 254, 446–449.

Nixon RA, Cataldo AM, Matthews PM. (2000) The endo-lysosomal system of neurons in Alzheimer's disease: a review. *Neurochem. Res.* 25, 1161–1172.

Noh JS, Kim EY, Kang JS, Kim HR, Oh YJ, Gwag BJ. (1999a) Neurotoxic and neuroprotective actions of catecholamines in cortical neurons. *Exp. Neurol.* 159, 217–224.

Noh KM, Lee JC, Ahn YH, Hong SH, Koh JY. (1999b) Insulin-induced oxidative neuronal injury in cortical culture: mediation by induced *N*-methyl-D-aspartate receptors. *Iubmb Life* 48, 263–269.

Novak G, Seeman P, Tallerico T. (2000) Schizophrenia: elevated mRNA for calcium-calmodulin-dependent protein kinase IIβ in frontal cortex. *Mol. Br. Res.* 82, 95–100.

Oakley RH, Laporte SA, Holt JA, Barak LS, Caron MG. (1999) Association of beta-arrestin with G protein-coupled receptors during clathrin-mediated endocytosis dictates the profile of receptor resensitization. *J. Biol. Chem.* 274, 32,248–32,257.

O'Brien RJ, Lau L-F, Huganir RL. (1998) Molecular mechanisms of glutamate receptor clustering at excitatory synapses. *Curr. Opin. Neurobiol.* 8, 364–369.

Ogimoto G, Yudowski GA, Barker CJ, Köhler M, Katz AI, Féraille E, Pedemonte CH, Berggren P-O, Bertorello M. (2000) G-protein-coupled receptors regulate Na$^+$, K$^+$-ATPase activity and endocytosis by modulating the recruitment of adaptor protein 2 and clathrin. *PNAS USA* 97, 3242–3247.

Ogita K, Enomoto R, Nakahara F, Ishitsubo N, Yoneda Y. (1995) A possible role of gluta-thione as an endogenous agonist at the *N*-methyl-D-aspartate recognition domain in rat brain. *J. Neurochem.* 64, 1088–1096.

Oh S-O, Hong J-H, Kim Y-R, Yoo H-S, Lee S-H, Lim K, Hwang B-D, Exton JH, Park S-K. (2000) Regulation of phospholipase D_2 by H_2O_2 in P12 cells. *J. Neurochem.* 75, 2445–2454.

Ohkuma S, Katsura M, Higo A, Shirotami K, Hara A, Tarumi C, Ohgi T. (2001) Peroxyni-trite affects Ca^2 inflow through voltage-dependent calcium channels. *J. Neurochem.* 76, 341–350.

Ohmori T, Abekawa T, Koyama T. (1996) The role of glutamate in behavioral and neuro-toxic effects of methamphetamine. *Neurochem. Int.* 29, 301–307.

Ohtsuka K, Suzuki T. (2000) Roles of molecular chaperones in the nervous system. *Br. Res. Bull.* 53, 141–146.

Okamoto M, Südhof TC. (1998) Mint 3: a ubiquitous mint isoform that does not bind to munc18-1 or -2. *Eur. J. Cell Biol.* 77, 161–165.

Ollinger K, Brunk UT. (1995) Cellular injury induced by oxidative stress is mediated through lysosomal damage. *Free Rad. Biol. Med.* 19, 565–574.

O'Neill GM, Fashena SJ, Golemis EA. (2000) Integrin signalling; a new Cas(t) of charac-ters enters the stage. *TICB* 10, 111–119.

Otake K, Ruggiero DA, Nakamura Y. (1995) Adrenergic innervation of forebrain neurons that project to the paraventricular thalamic nucleus in the rat. *Brain Res.* 657, 17–26.

Otmakhova NA, Otmakhov N, Mortenson LH, Lisman JE. (2000) Inhibition of the cAMP pathway decreases early long-term potentiation at CA1 hippocampal synapses. *J. Neurosci.* 20, 4446–4451.

Owen DJ, Wigge P, Vallis Y, Moore JD, Evans PR, McMahon HT. (1998) Crystal structure of the amphysib-2 SH3 domain and its role in the prevention of dynamin ring formation. *Embo J.* 17, 5273–5285.

Oyama Y, Noguchi S, Nakata M, Okada Y, Yamazaki Y, Funai M, Chikahisa L, Kanemaru K. (1999) Exposure of rat thymocytes to hydrogen peroxide increases annexin V binding to membranes: inhibitory actions of deferoxamine and quercetin. *Eur. J. Pharmacol.* 384, 47–52.

Pahl HL. (1999) Signal transduction from the endoplasmic reticulum to the cell nucleus. *Physiol. Rev.* 79, 683–701.

Pappone PA, Lee SC. (1996) Purinergic receptor stimulation increases membrane traf-ficking in brown adipocytes. *J. Gen. Physiol.* 108, 393–404.

Parton RG, Simons K, Dotti CG. (1992) Axonal and dendritic endocytic pathways in cultured neurons. *J. Cell Biol.* 119, 123–137.

Paulsen O, Sejnowski TJ. (2000) Natural patterns of activity and long-term synaptic plas-ticity. *Curr. Opin. Neurobiol.* 10, 172–179.

Pauly PC, Harris DA. (1998) Copper stimulates endocytosis of the prion protein. *J. Biol. Chem.* 273, 33107–33110.

Pearse BMF, Smith CJ, Owen DJ. (2000) Clathrin coat construction in endocytosis. *Curr. Opin. Struct. Biol.* 10, 220–228.

Pearson DL, Reimonenq RD, Pollard KM. (1999) Expression and purification of recom-binent mouse fibrillarin. *Prot. Express. Purif.* 17, 49–56.

Peitsch WK, Grund C, Kuhn C, Schnölzer M, Spring H, Schmelz M, Franke WW. (2000) Drebin is a widespread actin-associating protein enriched at junctional plaques, defining a specific microfilament anchorage system in polar epithelial cells. *Eur. J. Cell Biol.* 78, 767–778.

Perrone-Bizzozero NI, Sower AC, Bird ED, Benowitz LI, Ivins KJ, Neve RL. (1996) Levels of the growth-associated protein GAP-43 are selectively increased in association cortices in schizophrenia. *PNAS USA* 93, 14182–14187.

Petris MJ, Mercer JF. (1999) The Menkes protein (ATP7A; MNK) cycles via the plasma membrane both in basal and elevated extracellular copper using a C-terminal di-leucine endocytic signal. *Human Mol. Genetics* 8, 2107–2115.

Petrou C, Tashjian AH Jr. (1998) The thyrotropin-releasing hormone-receptor complex and G11α are both internalized into clathrin-coated vesicles. *Cell. Sig.* 10, 553–559.

Pfeffer SR. (1999) Motivating endosome motility. *Nature. Cell Biol.* 1, E145–E147.

Pickel VM, Nirenberg MJ, Milner TA. (1997) Ultrastructural view of central catechol-aminergic transmission: immunochemical localization of synthesizing enzymes, transporters and receptors. *J. Neurocytol.* 25, 843–856.

Pierce KL, Maudsley S, Daaka Y, Luttrell LM, Lefkowitz RJ. (2000a) Role of endocytosis in the activation of the extracellular signal-regulated kinase by sequestering and nonsequestering G protein-coupled receptors. *PNAS USA* 97, 1489–1494.

Pierce JP, van Leyen K, McCarthy JB. (2000b) Translocation machinery for synthesis of integral membrane and secretory proteins in dendritic spines. *Nature Neuroscience* 3, 311–313.

Pierini LM, Lawson MA, Eddy RJ, Hendey B, Maxfield FR. (2000) Oriented endocytic recycling of alpha5beta1 in motile neutrophils. *Blood*, 95, 2471–2480.

Poltorak M, Khoja I, Hemperly JJ, Williams JR, el-Mallakh R, Freed WJ. (1995) Distur-bances in cell recognition molecules (N-CAM and L1 antigen) in the CSF of pa-tients with schizophrenia. *Exp. Neurol.* 131, 266–272.

Poltorak M, Wright R, Hemperly JJ, Torrey EF, Issa F, Wyatt RJ, Freed WJ. (1997) Mono-zygotic twins discordant for schizophrenia are discordant for N-CAM and L1 in CSF. *Brain Res.* 751, 152–154.

Powell SK, Kleinman HK. (1997) Neuronal laminins and their cellular receptors. *Intern. J. Biochem. Cell Biol.* 291, 401–414.

Powell EM, Meiners S, DiProspero NA, Geller HM. (1997) Mechanisms of astrocyte-directed neurite guidance. *Cell Tissue Res.* 290, 385–393.

Prochiantz A. (2000) Messenger proteins: homeoproteins, TAT and others. *Curr. Opin. Cell Biol.* 12, 400–406.

Quartz SR, Sejnowski TJ. (1997) The neural basis of cognitive development: a constructiv-ist manifesto. *Behav. Brain Sci.* 20, 537–556.

Quinn CC, Gray GE, Hockfield S. (1999) A family of proteins implicated in axon guidance and outgrowth. *J. Neurobiol.* 41, 158–164.

Radulovic J, Blank T, Nijholt I, Kammermeier J, Speiss J. (2000) In vivo NMDA/dopa-mine interaction resulting in Fos production in the limbic system and basal gan-glia of the mouse brain. *Mol. Br. Res.* 75, 271–280.

Raimondi L, Banchelli G, Sgromo L. Pirisino R, Ner M, Parini A, Cambon C. (2000) Hy-drogen peroxide generation by monoamine oxidases in rat white adipocytes: role on cAMP production. *Eur. J. Pharmacol.* 395, 177–182.

Rajan I, Witte S, Cline HT. (1999) NMDA receptor activity stabilizes presynaptic retino-tectal axons and postsynaptic optic tectal cell dendrites in vivo. *J. Neurobiol.* 38, 357–368.

Ramsden JJ. (2000) MARCKS: a case of molecular exaptation? *Intern. J. Biochem. Cell. Biol.* 32, 475–479.

Raper JA. (2000) Semaphorins and their receptors in vertebrates and invertebrates. *Curr. Opin. Neurobiol.* 10, 88–94.

Raucher D, Sheetz MP. (1999) Membrane expansion increases endocytosis rate during mitosis. *J. Cell Biol.* 144, 497–506.

Rauhala P, Sziraki I, Chiueh CC. (1996) Peroxidation of brain lipids in vitro: nitric oxide versus hydroxyl radicles. *Free Rad Biol. Med.* 21, 391–399.

Ravati A, Ahlemeyer B, Becker A, Krieglstein J. (2000) Preconditioning-induced neurprotection is mediated by reactive oxygen species. *Br. Res.* 866, 23–32.

Rebec GV, Pierce RC. (1994) A vitamin as neuromodulator: ascorbate release into the extracellular fluid of the brain regulates dopaminergic and glutamatergic transmission. *Prog. Neurobiol.* 43, 537–565.

Rebec GV, Grabner CP, Johnson M, Pierce RC, Bardo MT. (1997) Transient increases in catecholaminergic activity in medial prefrontal cortex and nucleus accumbens shell during novelty. *Neuroscience* 76, 707–714.

Reid JM, O'Neil RG. (2000) Osmomechanical regulation of membrane trafficking in polarized cells. *Biochem. Biophys. Res. Comm.* 271, 429–434.

Reinheckel T, Ullrich O, Sitte N, Grune T. (2000) Differential impairment of 20S and 26S proteosome activities in human hematopoetic K562 cells during oxidative stress. *Arch. Biochem. Biophys.* 377, 65–68.

Renaudin A, Lehmann M, Girault J, McKerracher L. (1999) Organization of point contacts and neuronal growth cones. *J. Neurosci. Res.* 55, 458–471.

Rice ME, Cragg SJ, Greenfield SA. (1997) Characteristics of electrically evoked somatodendritic dopamine release in substantia nigra and ventral tegmental area in vitro. *J. Neurophysiol.* 77, 853–862.

Rico B, Cavada C. (1998) Adrenergic innervation of the monkey thalamus: an immuno-histochemical study. *Neuroscience* 84, 839–847.

Riedel G, Reymann KG. (1996) Metabotropic glutamate receptors in hippocampal long-term potentiation and learning and memory. *Acta Physiol. Scand.* 157, 1–19.

Righi M, Tongiorgi E, Cattaneo A. (2000) Brain-derived neurotrophic factor (BDNF) induces dendritic targeting of BDNF and tyrosine kinase B mRNAs in hippocampal neurons through a phosphatidylinositol-3 kinase-dependent pathway. *J. Neurosci.* 20, 3165–3174.

Ringstad N, Gad H, Löw P, Di Paolo G, Brodin L, Shupliakov O, De Camilli P. (1999) Endophilin/SH3p4 is required for the transition from early to late stages in clathrin-mediated synaptic vesicle endocytosis. *Neuron* 24, 143–154.

Rioult-Pedotti M-S, Freidman D, Donoghue JP. (2000) Learning-induced LTP in neocortex. *Science* 290, 535–536.

Rocha M, Sur M. (1995) Rapid acquisition of dendritic spines by visual thalamic neurons after blockade of N-methyl-D-aspartate receptors. *PNAS USA* 92, 8026–8030.

Roche KW, Tu JC, Petralia RS, Xiao B, Wenthold RJ, Worley PF. (1999) Homer 1b regulates the trafficking of group I metabotropic glutamate receptors. *J. Biol. Chem.* 274, 25,953–25,957.

Rockwell P, Yuan H, Magnusson R, Figueiredo-Pereira ME. (2000) Proteosome inhibition in neuronal cells induces a proinflammatory response manifested by upregulation of cyclooxygenase-2, its accumulation as ubiquitin conjugates and production of prostaglandin PGE (2). *Arch. Biochem. Biophys.* 374, 325–333.

Rodriguez MA, Pesold C, Liu WS, Kriho V, Guidotti A, Pappas GD, Costa E. (2000) Colocalization of integrin receptors and reelin in dendritic spine postsynaptic densities of adult non-human primate cortex. *PNAS USA* 97, 3550–3555.

Roos J, Hummel T, Ng N, Klämbt C, Davis GW. (2000) *Drosophila* futsch regulates synaptic microtubule organization and is necessary for synaptic growth. *Neuron* 26, 371–382.

Roseberry AG, Hosey MM. (1999) Trafficking of M(2) muscarinic acetylcholine receptors. *J. Biol. Chem.* 274, 33,671–33,676.

Rosenberg PA, Li Y, Ali S, Altiok N, Back SA, Volpe JJ. (1999) Intracellular redox state determines whether nitric oxide is toxic or protective to rat oligodendrocytes in culture. *J. Neurochem.* 73, 476–484.

Rosenthal JA, Chen H, Slepnev VI, Pellegrini L, Salcini AE, Di Fiore PP, Di Camilli P. (1999) The epsins define a family of proteins that interact with components of the clathrin coat and contain a new protein module. *J. Biol. Chem.* 274, 33,959–33,965.

Rossi DJ, Hamann M. (1998) Spillover-mediated transmission at inhibitory synapses promoted by high affinity alpha6 subunit GABA (A) receptors and glomerular geometry. *Neuron* 20, 783–795.

Roth BL, Willins DL. (1999) What's all the RAVE about receptor internalization? *Neuron* 23, 629–631.

Rotin D, Staub O, Haguenauer-Tsapis R. (2000) Ubiquination and endocytosis of plasma membrane proteins: role of Nedd4/Rsp5p family of ubiqutin-protein ligases. *J. Memb. Biol.* 176, 1–17.

Rottkamp CA, Raina AK, Zhu X, Gaieu E, Bush AI, Atwood CS, Chevion M, Perry G, Smith MA. (2001) Redox-active iron mediates amyloid-β toxicity. *Free Rad. Biol. Med.* 30, 447–450.

Rouze NC, Schwartz EA. (1998) Continuous and transient vesicle cycling at a ribbon synapse. *J. Neurosci.* 18, 8614–8624.

Roy AM, Parker JS, Parrish CR, Whittaker GR. (2000) Early stages of influenza virus entry into Mv-1 lung cells: involvement of dynamin. *Virology* 267, 17–28.

Rumpel S, Hatt H, Gottmann K. (1998) Silent synapses in the developing rat visual cortex: evidence for postsynaptic expression of synaptic plasticity. *J. Neurosci.* 18, 8863–8874.

Rusakov DA, Kullmann DM. (1998) Extrasynaptic glutamate diffusion in the hippocampus: ultrastructural constraints, uptake, and receptor activation. *J. Neurosci* 18, 3158–3170.

Rutherford LC, Nelson SB, Turrigiano GG. (1998) BDNF has opposite effects on the quantal amplitude of pyramidal neuron and interneuron excitatory synapses. *Neuron* 21, 521–530.

Rutledge LT, Wright C, Duncan J. (1974) Morphological changes in pyramidal cells in mammalian neocortex associated with increased use. *Exp. Neurol.* 44, 209–228.

Rybnikova E, Damdimopoulos AE, Gustafsson J-A, Spyrou G, Pelto-Huikko M. (2000) Expression of novel antioxidant thioredoxin-2 in the rat brain. *Eur. J. Neurosci.* 12, 1669–1678.

Sabatini BL, Maravall M, Svoboda K. (2001) Ca^{2+} signaling in dendritic spines. *Curr. Opin. Neurobiol.* 11, 349–356.

Sah R, Schwartz-Bloom RD. (1999) Optical imaging reveals elevated intracellular chloride in hippocampal pyramidal neurons after oxidative stress. *J. Neurosci.* 19, 9209–0217.

Sakai T, Furuyama T, Ohoka Y, Miyazaki N, Fujioka S, Sugimoto H, Amasaki M, Hattori S, Matsuya T, Inagaki S. (1999) Mouse semaphorin H induces PC12 neurite outgrowth activating Ras-mitogen-activated protein kinase signaling pathway via Ca^{2+} influx. *J. Biol. Chem.* 274, 29666–29671.

Sakamoto K, Yamasaki Y, Kaneto H, Fujitani Y, Matsuoka T. Yoshioka R, Tagawa T, Matsohisa M, Kajimoto Y, Hori M. (1999) Identification of oxidative stress-regulated genes in rat aortic smooth muscle cells by suppression subtractive hybridization. *FEBS Lett,* 461, 47–51.

Sala C, Rudolph-Correia S, Sheng M. (2000) Developmentally regulated NMDA receptor-dependent dephosphorylation of cAMP response element-binding protein (CRAB) in hippocampal neurons. *J. Neurosci.* 20, 3529–3536.

Salinas PC. (1999) Wnt factors in axonal remodelling and synaptogenesis. *Biochem. Soc. Symp.* 65, 101–109.

Salvemini D, Misko TP, Masferrer JL, Seibert K, Currie MG, Needleman P. (1993) Nitric oxide activates cyclooxygenase enzymes. *PNAS USA* 90, 7240–7244.

Samanta S, Perkinton MS, Morgan M, Williams RJ. (1998) Hydrogen peroxide enhances signal-responsive arachidonic acid release from neurons: role of mitogen-activated protein kinase. *J. Neurochem.* 70, 2082–2090.

Sampath D, Perez-Polo R. (1997) Regulation of antioxidant enzyme expression by NGF. *Neurochem. Res.* 22, 351–362.

Sassoè-Pognetto, M Fritschy, JM. (2000) Mini-review: gephyrin, a major postsynaptic protein of GABAergic synapses. *Eur. J. Neuroscience*, 12, 2205–2210.

Sassoè-Pugnetto M, Cantino D, Panzanelli P, Verdun-di-Cantogno L, Giustetto M, Margois F, de Biasi L, Fasolo A. (1993) Presynaptic co-localization of carnosine and glutamate in olfactory neurons. *NeuroReport* 5, 7–10.

Satoh-Horikawa K, Nakanishi H, Takahashi K, Miyahara M, Nishimura M, Tachibana K, Mizoguchi A, Takai Y. (2000) Nectin-3, a new member of immunoglobulin-like cell adhesion molecules that shows homophilic and heterophilic cell-cell adhesion activities. *J. Biol. Chem.* 275, 10, 291–10,299.

Sattler R, Charlton MP, Hafner M, Tymianski M. (1998) Distinct influx pathways, not calcium load, determines neuronal vulnerability to calcium neurotoxicity. *J. Neurochem.* 71, 2349–2364.

Sawada H, Ibi M, Kihara T, Urushitani M, Akaide A, Kimura J, Shimohama S. (1998) Dopamine D2-type agonists protect mesencephalic neurons from glutamate neurotoxicity: mechanisms of neuroprotective treatment against oxidative stress. *Ann. Neurol.* 44, 110–119.

Schaefer AW, Kamiguchi H, Wong EV, Beach CM, Landreth G, Lemmon V. (1999) Activation of the MAPK signal cascade by the neural cell adhesion molecule L1 requires L1 internalization. *J. Biol. Chem.* 274, 37,965–37,973.

Schauwecker PE, McNeill TH. (1996) Dendritic remodeling of dentate granule cells following a combined entorhinal cortex / fimbria fornix lesion. *Exp. Neurol.* 141, 145–153.

Scheetz AJ, Nairn AC, Constantine-Paton M. (2000) NMDA receptor-mediated control of protein synthesis at developing synapses. *Nature Neuroscience* 3, 211–216.

Scheiffele P, Fan J, Choih J, Fetter R, Serafini T. (2000) Neuroligin expressed in nonneuronal cells triggers presynaptic development in contacting axons. *Cell* 101, 657–669.

Scherer SS. (1999) Nodes, paranodes, and incisures: from form to function. *Ann. N.Y. Acad. Sci.* 883, 131–142.

Schiavo G, Matteoli M, Montecucco C. (2000) Neurotoxins affecting neuroexocytosis. *Physiol. Rev.* 80, 717–766.

Schinkmann KA, Kim T-A, Avraham S. (2000) Glutamate-stimulated activation of DNA synthesis via mitogen-activated protein kinase in primary astrocytes: involvement of protein kinase C and related adhesion focal tyrosine kinase. *J. Neurochem.* 74, 1931–1940.

Schmid R-S, Pruitt WM, Maness PF. (2000) A MAPK kinase-signaling pathway mediates neurite outgrowth on L1 and requires Src-dependent endocytosis. *J. Neurosci.* 20, 4177–4188.

Schultz W. (1997) Dopamine neurons and their role in reward mechanisms. *Curr. Opin. Neurobiol.* 7, 191–197.

Schultz W. (1998) Predictive reward signal of dopamine neurons. *J. Neurophysiol.* 80, 1–27.

Schulz JB, Matthews RT, Klockgether T, Dichgans J, Beal MF. (1997) The role of mitochondrial dysfunction and neuronal nitric oxide in animal models of neurodegenerative diseases. *Mol. Cell. Biochem.* 174, 193–197.

Schuman EM. (1997) Synapse specificity in long-term information storage. *Neuron* 18, 339–342.

Schuman EM (1999) mRNA trafficking and local protein synthesis at the synapse. *Neuron* 23, 645–648.

Schwartz AL. (1995) Receptor cell biology: receptor-mediated endocytosis. *Pediat. Res.* 38, 835–843.

Schwartz MA, Shattil SJ. (2000) Signaling networks linking integrins and rho family GTPases. *TIBS* 25, 388–391.

Schwartz BE, Sem-Jacobsen C, Petersen MC. (1956) Effects of mescaline, LSD-25 and adrenochrome on depth electrograms in man. *Arch. Neurol. Psychiat.* 75, 579–587.

Schwencke C, Okomura S, Yamamoto M, Geng Y-J, Ishikawa Y. (1995) Colocalization of β-adrenergic receptors and caveolin within plasma membrane. *J. Cell. Biochem.* 75, 64–72.

Scott EK, Luo L. (2001) How dendrites take their shape? Nature Neuroscience. 4, 359–365.

Sen CK. (1998) Redox signaling and the emerging therapeutic potential of thiol antioxidants. *Biochem. Pharmacol.* 55, 1747–1751.

Sengar AS, Wang W, Bishay J, Cohen S, Egan SE. (1999) The EH and SH3 domains of Ese proteins regulate endocytosis by linking to dynamin and Eps15. *Embo J.* 18, 1159–1171.

Sermasi E, Margotti E, Cattaneo A, Domenici L. (2000) TrkB signalling controls LTP but not LTD expression in the developing rat visual cortex. *Eur. J. Neurosci.* 12, 1411–1419.

Servitja J-M, Masgrau R, Pardo R, Sarri E, Picatoste F. (2000) Effects of oxidative stress on phospholipid signaling in rat cultured astrocytes and brain slices. *J. Neurochem.* 75, 788–794.

Sesack SR, Pickel VM. (1990) In the rat nucleus accumbens, hippocampal and catecholaminergic terminals converge on spiny neurons and are in apposition to each other. *Brain Res.* 527, 266–279.

Sesack SR, Hawrylak VA, Matus C, Guido MA, Levey AI. (1998a) Dopamine axon varicosities in the prelimbic division of the rat prefrontal cortex exhibit sparse immunoreactivity for the dopamine transporter. *J. Neurosci.* 18, 2697–2708.

Sesack SR, Hawrylak VA, Melchitsky S, Lewis DA. (1998b) Dopamine innervation of a subclass of local circuit neurons in monkey prefrontal cortex: ultrastructural analysis of tyrosine hydroxylase and parvalbumin immunoreactive structures. *Cerebral Cortex,* 8, 614–622.

Shaw PJ, Salt TE. (1997) Modulation of sensory and excitatory amino acid responses by nitric oxide donors and glutathione in the ventrobasal thalamus of the rat. *Eur. J. Neurosci.* 9, 1507–1513.

Shen X-M, Dryhurst G. (1996) Oxidation chemistry of (−)-norepinephrine in the presence of L-cysteine *J. Med. Chem.* 39, 2018–2029.

Shen K, Teruel MN, Connor JH, Shenolikar S, Meyer T. (2000) Molecular memory by reversible translocation of calcium/calmodulin-dependent protein kinase II. *Nature Neuroscience* 3, 881–886.

Sheng JG, Mrak RE, Bales KR, Cordell B, Paul SM, Jones RA, Woodward S, Zhou XQ, McGinness JM, Griffin WS. (2000) Overexpression of the neuritotrophic cytokine S100beta precedes the appearance of neuritic beta-amyloid plaques in APPV717F mice. *J. Neurochem.* 74, 295–301.

Sheng M, Lee SH. (2001) AMPA receptor trafficking and the control of synaptic transmission. *Cell,* 105, 825–828.

Shi SH, Hayashi Y, Petralia RS, Zaman SH, Wenthold RJ, Svoboda K, Malinow R. (1999) Rapid spine delivery and redistribution of AMPA receptors after synaptic NMDA receptor activation. *Science* 284, 1811–1816.

Shimada A, Mason CA, Morrison ME. (1998) TrkB signaling modulates spine density and morphology independent of dendrite structure in cultured neonatal Purkinjie cells. *J. Neurosci.* 18, 8559–8570.

Shiosaka S, Yoshida S. (2000) Synaptic microenvironments—structural plasticity, adhesion molecules, proteases and their inhibitors. *Neurosci. Res.* 237, 385–389.

Shiraisni Y, Mizutani A, Bito H, Fujisawa K, Narumiya S, Mikoshiba K, Furuichi T. (1999) Cupidin, an isoform of Homer/Vesl, interacts with the actin cytoskeleton and activated rho family small GTPases and is expressed in developing mouse cerebellar granule cells. *J. Neurosci.* 19, 8389–8400.

Shirvan A, Ziv I, Fleminger G, Shina R, He Z, Brudo I, Melamed E, Barzilai A. (1999) Semaphorins as mediators of neuronal apoptosis. *J. Neurochem.* 73, 961–971.

Shors TJ, Elkabes S, Selcher JC, Black IB. (1997) Stress persistently increases NMDA receptor-mediated binding of [^{3}H]PDBu (a marker for protein kinase C) in the amygdala, and re-exposure to the stressful context reactivates the increase. *Br. Res.* 750, 293–300.

Shors TJ, Matzel LD. (1997) Long-term potentiation: what's learning got to do with it? *Behav. Brain Sci.* 20, 597–614.

Shpetner H, Joly M, Hartley D, Corvera S. (1996) Potential sites of PI-3 kinase function in the endocytic pathway revealed by the PI-3 kinase inhibitor, wortmannin. *J. Cell Biol.* 132, 595–605.

Sidhu A. (1999) Oxidative stress and neuronal signaling. *J. Neurochem.* 72, S2.

Sik A, Gulácsi A, Lai Y, Doyle WK, Pacia S, Mody I, Freund TF. (2000) Localization of the A kinase anchoring protein AKAP79 in the human hippocampus. *Eur. J. Neurosci.* 12, 1155–1164.

Simbürger E, Plaschke M, Kirsch J, Nitsch R. (2000) Distribution of the receptor-anchoring protein gephyrin in the rat dendate gyrus and changes following entorhinal cortex lesion. *Cereb. Cortex* 10, 422–432.

Simpson GLW, Ortwerth BJ. (2000) The non-oxidative degradation of ascorbic acid at physiological conditions. *Biochem. Biophys. Acta* 1501, 12–24.

Simpson F, Hussain NK, Qualmann B, Kelly RB, Kay BK, McPherson PS, Schmid SL. (1999) SH3-domain-containing proteins function as distinct steps in clathrin-coated vesicle function. *Nature Cell Biol.* 1, 119–124.

Siraki AG, Smythies J, O'Brien PJ (2000) Superoxide radical scavenging and attenuation of hypoxia-reoxygenation injury by neurotransmitter ferric complexes in isolated rat hepatocytes. *Neurosci. Lett.* 296, 37–40

Skarpen E, Johannessen LE, Bjerk K, Fasteng H, Guren TK, Lindeman B, Thoresen GH, Christoffersen T, Stang E, Huitfeldt HS, Madshus IH. (1998) Endocytosed epidermal growth factor (EGF) receptors contribute to the EGF-mediated growth arrest in A431 cells by inducing a sustained increase in p21/CIP1. *Exp. Cell Res.* 243, 161–172.

Skeberdis VA, Lan J-Y, Zheng X, Zukin RS, Bennett MUL (2001) Insulin promotes rapid delivery of N-methyl-D-aspartate receptors to the cell surface by exocytosis. *PNAS USA* 98, 3561–3568.

Slepnev VI, Ochoa GC, Butler MH, Grabs D, Camilli PD. (1998) Role of phosporylation in regulation of the assembly of endocytic coat complexes. *Science* 281, 821–824.

Smalheiser NR, Dissanayake S, Kapil A. (1996) Rapid regulation of neurite outgrowth and retraction by phospholipse A2-derived arachidonic acid and its metabolites. *Brain Res.* 721, 39–48.

Smiley JF, Levey AI, Ciliax BJ, Goldman-Rakic PS. (1994) D1 dopamine receptor immunoreactivity in human and monkey cerebral cortex: predominant and extrasynaptic localization in dendritic spines. *PNAS USA* 91, 5720–5724.

Smith RM, Baibakov B, Ikebuchi Y, White BH, Lambert NA, Kaczmarek LK, Vogel SS. (1999) Exocytic insertion of calcium channels constrains compensatory endocytosis to sites of exocytosis. J. Cell Biol. 148, 755–767.

Smith MA, Rottkamp CA, Nunomura A, Raina AK, Perry G. (2000) Oxidative stress in Alzheimer's disease. *Biochim. Biophys. Acta* 1502, 139–144.

Smythies J. (1997) The biochemical basis of synaptic plasticity and neurocomputation; a new theory. *Proc. R. Soc. Lond. B.* 264, 575–579.

Smythies J. (1999) The neurotoxicity of glutamate, dopamine, iron and reactive oxygen species: Functional interrelationships in health and disease: A review-discussion. *Neurotox. Res.* 1, 27–37.

Smythies J. (2000) What is the function of receptor and membrane endocytosis at the postsynaptic neuron? *Proc. R. Soc. Lond B.* 267, 1363–1367.

Smythies JR, Gottfries C-G, Regland B. (1997) Disturbances of one-carbon metabolism in neuropsychiatric disorders; a review. *Biol. Psychiat.* 41, 230–233.

Snyder GL, Allen PB, Fienberg AA, Valle CG, Huganin RL, Nairn AC, Greengard P. (2000) Regulation by phosphorylation of the gluR1 AMPA receptor in the neostriatum by dopamine and psychostimulants in vivo. *J. Neurosci.* 20, 4480–4488.

Sojakka K, Punnonen EL, Majomäki VS. (1999) Isoprotenerol inhibits fluid-phase endocytosis from early to late endosomes. *Eur. J. Cell Biol.* 78, 161–169.

Song JH, Shin SH, Ross GM. (1999) Prooxidant effects of ascorbate in rat brain slices. *J. Neurosci. Res.* 58, 328–336.

Sönnichsen B, De Renzis S, Nielsen E, Rietdof J, Zerial M, (2000) Distinct membrane domains on endosomes in the recycling pathway visualized by multicolor imaging of Rab4, Rab5 and Rab11. *J. Cell Biol.* 149, 901–914.

Sorensen SD, Linseman DA, McEwen EL, Heacock AM, Fisher SK. (1998) A role for a wortmannin-sensitive phosphatidylinositol-4-kinase in the endocytosis of muscarinic cholinergic receptors. *Mol. Pharmacol.* 53, 827–836.

Spacek J, Harris KM. (1997) Three-dimensional organization of smooth endoplasmic reticulum in hippocampal CA1 dendrites and dendritic spines of the immature and mature rat. *J. Neurosci.* 17, 190–203.

Spanagel R, Weiss F. (1999) The dopamine hypothesis of reward: past and current status. *TINS* 22, 521–527.

Spiro DJ, Boll W, Kirchhausen T, Wessling-Resnick M. (1996) Wortmannin alters the transferrin receptor endocytic pathway in vivo and in vitro. *Mol. Biol. Cell* 7, 355–367.

Stahl B, Tobaben S, Südhof TC. (1999) Two distinct domains in hsc70 are essential for the interaction with the synaptic vesicle cysteine string protein. *Eur. J. Cell Biol.* 78, 375–381.

Stenmark H. (2000) Cycling lipids. *Curr. Biol.* 10, R57–R59.

Stern JE, Armstrong WE. (1998) Reorganization of the dendritic trees of oxytocin and vasopressin neurons of the rat supraoptic nucleus during lactation. *J. Neurosci.* 18, 841–853.

Sterner-Kock A, Braun RK, van der Vliet A, Schrenzel MD, McDonald RJ, Kabbur MB, Vulliet PR, Hyde DM. (1999) Substance P primes the formation of hydrogen peroxide and nitric oxide in human neutrophils. *J. Leukocyte Biol.* 65, 834–840.

Stevens GR, Zhang C, Berg MM, Lambert MP, Barber K, Cantallops I, Routtenberg A, Klein WL. (1996) CNS neuronal focal adhesion kinase forms clusters that co-localize with vinculin. *J. Neurosci. Res.* 46, 445–455.

Stoeckli ET, Ziegler U, Bleiker AJ, Groscurth P, Sonderegger P. (1996) Clustering and functional cooperation of Ng-CAM and axonin-1 in the substratum-contact area of growth cones. *Develop. Biol.* 177, 15–29.

Stone R, Stewart VO, Hurst RD, Clark JB, Heales SJR. (1999) Astrocytes release and preserve antioxidants: implications for neuroprotection. *Biochem. Soc. Trans.* 27, A152.

Strous GJ, Govers R. (1999) The ubiquitin-proteosome system and endocytosis. *J. Cell Biol.* 112, 1417–1423.

Subtil A, Hémar A, Dautry-Varsat A. (1994) Rapid endocytosis of interleukin 2 receptors when clathrin-coated pit endocytosis is inhibited. *J. Cell Sci.* 107, 3461–3468.

Sugahara M, Shiraishi H. (1999) Dopamine D1 and D2 receptor agents and their interaction influence the synaptic density of the rat prefrontal cortex. *Neurosci. Lett.* 259, 141–144.

Sulzer D, Joyce MP, Lin L, Geldwert D, Haber SN, Hattori T Rayport S. (1998) Dopamine neurons make glutamatergic synapses in vitro. *J. Neurosci.* 18, 4588–4602.

Suzuki Y, Ono Y. (1999) Involvement of reactive oxygen species produced via NADPH oxidase in tyrosine phosphorylation in human B- and T-lineage lymphoid cells. *Biochem. Biophys. Res. Com.* 255, 262–267.

Suzuki YJ, Forman HJ, Sevanian A. (1997) Oxidants as stimulators of signal transduction. *Free Rad. Biol. Med.* 22, 269–285.

Sweatt JD. (2000) The neuronal MAP kinase cascade: a biochemical signal integration system subserving synaptic plasticity and memory. *J. Neurochem.* 76, 1–10.

Sydor AM, Su AL, Wang FS, Xu A, Jay DG. (1996) Talin and vinculin play distinct roles in filopodial motility in the neuronal growth cone. *J. Cell Biol.* 134, 1197–1207.

Szabó C. (1996) Physiological and pathophysiological roles of nitric oxide in the central nervous system. *Br. Res. Bull.* 41, 131–141.

Szapiro G, Izquierdo LA, Alonso D, Barros D, Paratcha G, Ardenghi G, Periera P, Medina JH, Izquierdo I. (2000) Participation of hippocampal metabotropic glutamate receptors, protein kinase A and mitogenic-activated protein kinases in memory retrieval. *Neuroscience* 99, 1–5.

Szekeres PG, Koenig JA, Edwardson JM. (1998) The relationship between agonist intrinsic activity and the rate of endocytosis of muscarinic receptors in a human neuroblastoma cell line. *Mol. Pharmacol.* 53, 759–765.

Tadokoro S, Tachibana T, Imanaka T, Nishida W, Sobue K. (1999) Involvement of unique leucine-zipper motif of PSD-Zip45 (Homer1c/vesl-1L) in group 1 metabotropic glutamate receptor clustering. *PNAS USA* 96, 13,801–13,806.

Tanaka H, Shan W, Phillips GR, Arndt K, Bozdagi O, Shapiro L, Huntley GW, Benson DL, Colman DR. (2000) Molecular modification of N-cadherin in response to synaptic activity. *Neuron* 25, 93–107.

Tanaka K, Miyazaki I, Fujita N, Haque ME, Asanuma M, Ogawa N. (2001) Molecular mechanisms of activation of glutathione system by ropinirole, a selective dopamine D2 agonist. *Neurochem. Res.* 26, 31–36.

Taubman G, Jantz H. (1957) Untersuchung über die dem adrenochrom zugeschrieben psychotoxischen wirkungen. *Nervenartz* 28, 485–488.

Tcherepanov AA, Sokolov BP. (1997) Age-related abnormalities in expression of mRNAs encoding synapsin 1A, synapsin 1B, and synaptophysin in the temporal cortex of schizophrenics. *J. Neurosci. Res.* 49, 639–644.

Terry-Lorenzo RT, Inoue M, Connor JH, Haystead TA, Armbruster BN, Gupta RP, Oliver CJ, Shenolikar S. (2000) Neurofilament-L is a protein phosphatase-1-binding protein associated with neuronal plasma membrane and post-synaptic density. *J. Biol. Chem.* 275, 2439–2446.

Thiels E, Urban NN, Gonzales-Burgos GR, Kanterewicz B, Barrionuevo G, Chu CT, Oury TD, Klann E. (2000) Impairment of long-term potentiation and associative memory in mice that overexpress extracellular superoxide dismutase. *J. Neurosci.* 20, 7631–7639.

Thome J, Sakai N, Shin K-H, Steffen C, Zhang Y-J, Impey S, Storm D, Duman RS. (2000) cAMP response element-mediated gene transcription is upregulated by chronic antidepressant treatment. *J. Neurosci.* 20, 4030–4036.

Thompson PM, Sower AC, Perrone-Bizzozero NI. (1998) Altered levels of the synaptosomal associated protein SNAP-25 in schizophrenia. *Biol. Psychiat.* 43, 239–243.

Tokuda M, Hatase O. (1998) Regulation of neuronal plasticity in the central nervous system by phosphorylation and dephosphorylation. *Mol. Neurobiol.* 17, 137–156.

Tong G, Shepherd D, Jahr CE. (1995) Synaptic desensitization of NMDA receptors by calcineurin. *Science* 267, 1510–1512.

Torreilles F, Salman-Tabcheh S, Guérin M-C, Torreilles J. (1999) Neurodegenerative disorders; the role of peroxynitrite. *Br. Res. Rev.* 30, 153–163.

Tramontin AD, Brenowitz EA. (2000) Seasonal plasticity in the adult brain. *TINS* 23, 251–258.

Trapaidze N, Gomes I, Cvejic S, Bansinath M, Devi LA. (2000) Opioid receptor endocytosis and activation of MAP kinase pathway. *Mol. Br. Res.* 76, 220–228.

Trejo J, Coughlin SR. (1999) The cytoplasmic tails of protease-activated receptor-1 and substance P receptor specify sorting to lysosomes versus recycling. *J. Biol. Chem.* 274, 2216–2224.

Trischler M. Stoorvogel W, Ullrich O. (1999) Biochemical analysis of distinct Rab5- and Rab11-positive endosomes along the transferrin pathway. *J. Cell Sci.* 112, 4773–4783.

Trombley PQ, Horning MS, Blakemore LJ. (1998) Carnosine modulates zinc and copper effects on amino acid receptors and synaptic transmission. *NeuroReport* 9, 3503–3507.

Trotti D, Volterra A, Lehre KP, Rossi D, Gjesdal O, Racagni G, Danbolt NC. (1995) Arachidonic acid inhibits a purified and reconstituted glutamate transporter directly from the water phase and not via the phospolipid membrane. *J. Biol. Chem.* 270, 9890–9895.

Trotti D, Rossi D, Gjesdal O, Levy LM, Racagni G, Danbolt NC, Volterra A. (1996) Peroxynitrite inhibits glutamate transporter subtypes. *J. Biol. Chem.* 271, 5976–5979.

Tu JC, Xiao B, Naisbitt S, Yuan JP, Petralia RS, Brakeman P, Doan A, Aakalu VK, Lanahan AA, Sheng M, Worley PF. (1999) Coupling of mGluR/Homer and PSD-95 complexes by the Shank family of postsynaptic density proteins. *Neuron* 23, 583–592.

Turrigiano GG. (2000) AMPA receptors unbound: membrane cycling and synaptic plasticity. *Neuron* 26, 5–8.

Turrigiano GG, Nelson SB. (1998) Thinking globally, acting locally; AMPA receptor turnover and synaptic strength. *Neuron* 21, 933–935.

Ullrich V, Bachschmid M. (2000) Superoxide as a messenger of endothelial function. *Biophys. Biochem. Res. Comm.* 278. 1–8.

Undie AS, Berki AC, Beardsley K. (2000) Dopaminergic behaviors and signal transduction mediated through adenylate cyclase and phospholipase C pathways. *Neuropharmacology* 39, 75–87.

Uylings HBM, Kuypers K, Diamond MC, Veltman WAM. (1978) Effects of differential environments on plasticity of dendrites of cortical pyramidal neurons in adult rats. *Exp. Neurol.* 62, 658–677.

Valetti C, Wetzel DM, Schrader M, Hasbani MJ, Gill SR, Kreis TE, Schroer TA. (1999) Role of dynamin in endocytic traffic; effects of dynamin overexpression and colocalization with CLIP-170. *Mol. Biol.Cell* 10, 4107–4120.

Valverde F. (1971) Rate and extent of recovery from dark rearing in the visual corex of the mouse. *Brain Res.* 33, 1–11.

van Aelst L, D'Souza-Schorey C. (1997) Rho GTPases and signaling networks. *Genes Dev.* 11, 2295–2322.

Vancura KL, Jay DG. (2000) G proteins and axon growth. *Sem. neurosci.* 9, 209–219.

van Gassen G, Annaert W, van Broeckhoven C. (2000) Binding partners of Alzheimer's disease proteins: are they physiologically relevant? *Neurobiol. Dis.* 7, 135–151.

van Voorst F, de Kruiff B. (2000) Role of lipids in the translocation of proteins across membrane. *Biochem. J.* 347, 601–612.

Varga V, Jenei Z, Janáky R, Saransaari P, Oja SS (1997) Glutathione is an endogenous ligand of rat brain N-methyl-D-aspartate (NMDA) and 2-amino-3-hydroxy-5-methyl-4-isoxazoleproprionate (AMPA) receptors. *Neurochem. Res.* 22, 1165–1171.

Vawter MP, Cannon-Spoor HE, Hemperly JJ, Hyde TM, VanderPutten DM, Kleinman JE, Freed WJ. (1998a) Abnormal expression of cell recognition molecules in schizophrenia. *Exp. Neurol.* 149, 424–432.

Vawter MP, Hemperley JJ. Freed WJ, Garver DL. (1998b) CSF N-CAM in neuroleptic-naïve first-episode patients with schizophrenia. *Schiz. Res.* 34, 123–131.

Vawter MP, Frye MA, Hemperly JJ, VanderPutten DM, Usen DM, Doherty P, Saffell JL, Issa F, Post RM, Wyatt RJ, Freed WJ. (2000) Elevated concentration of N-CAM VASE isoforms in schizophrenia. *J. Psychiatr. Res.* 34, 25–34.

Ventura R, Harris KM. (1999) Three-dimensional relationships between hippocampal synapses and astrocytes. *J. Neurosci,* 19, 6897–6906.

Verona M, Zanotti S, Schäfer T, Racagni G, Popoli M. (2000) Changes of synaptotagmin interaction with t-SNARE proteins in vitro after calcium/calmodulin-dependent phosphorylation. *J. Neurochem.* 74, 209–221.

Vesce S, Bezzi P, Volterra A. (1999) The active role of astrocytes in synaptic transmission. *Cell. Mol. Life Sci.* 56, 991–1000.

Vickery RG, von Zastrow CC. (1998) Subtype specific differences in dopamine receptor endocytosis. *Abst. Soc. Neurosci.* 24, 23.

Vieira AV, Lamaze C, Schmid SL (1996) Control of EGF receptor signaling by clathrin-mediated endocytosis. *Science* 274, 2086–2089.

Vinadé L, Dosemeci A. (2000) Regulation of the phosphorylation state of the AMPA receptor GluR1 subunit in the postsynaptic density. *Cell. Mol. Neurobiol.* 20, 451–463.

Vögler O, Nolte B, Voss M, Schmidt M, Jakobs KH, van Koppen CJ. (1999) Regulation of muscarinic acetylcholine receptor sequestration and function by beta-arrestin. *J. Biol. Chem.* 274, 12333–12338.

Volkmar FR, Greenough WT. (1972) Rearing complexity affects branching of dendrites in the visual cortex of the rat. *Science* 176, 1445–1447.

Volkmer H, Zacharias U, Nörenberg U, Rathjen FG. (1998) Dissection of complex molecular interactions of neurofascin with axonin-1, F11, and tenascin-R, which promote attachment and neurite formation of tectal cells. *J. Cell Biol.* 24, 1083–1093.

Vrecl M, Heding A, Hanyaloglu A, Taylor PL, Eidne KA. (2000) Internalization kinetics of the gonadotrophin-releasing hormone (GnRH) receptor. *Eur. J. Physiol.* 439S, R19–R20.

Walikonis RS, Jensen ON, Mann M, Provance DW Jr., Mercer JA, Kennedy MB. (2000) Identification of proteins in the postsynaptic density fraction by mass spectrometry. *J. Neurosci.* 20, 4069–4080.

Wallace CS, Kilman VL, Withers GS, Greenough WT. (1992) Increases in dendritic length in occipital cortex after 4 days of differential housing in weanling rats. *Behav. Neural Biol.* 58, 64–68.

Walz R, Roesler R, Quevedo J, Sant'Anna MK, Madruga M, Rodrigues C, Gottfried C, Medina JH, Izquierdo I. (2000) Time-dependent impairment of inhibitory avoidance retention in rats by posttraining infusion of a mitogen-activated protein kinase inhibitor into cortical and limbic structures. *Neurobiol. Learn. Memory* 73, 11–20.

Wan HI, DiAntonio A, Fetter RD, Bergstrom K, Strauss R, Goodman CS. (2000) Highwire regulates synaptic growth in *Drosophila. Neuron* 26, 313–329.

Wang Y, Floor E. (1998) Hydrogen peroxide inhibits the vacuolar H^+-ATPase in brain synaptic vesicles at micromolar concentrations. *J. Neurochem.* 70, 646–652.

Wang H-Y, Lee DHS, Davis CB, Shank RP. (2000) Amygdaloid $A\beta_{1-42}$ binds selectively and with picomolar affinity to $\alpha7$ nicotinic acetylcholine receptors. *J. Neurochem.* 75, 1155–1161.

Weingarten P, Zhou Q-Y. (2001) Protection of intracellular dopamine cytoxicity by dopamine disposition and metabolism factors. *J. Neurochem.* 77, 776–785.

Wheal HV, Chen Y, Mitchell J, Schachner W, Maerz W, Wieland H, van Rossum D, Kirsch J. (1998) Molecular mechanisms that underlie structural and functional chnages at the postsynaptic membrane during synaptic plasticity. *Prog. Neurobiol.* 55, 611–640.

Whistler JL, von Zastrow M. (1999) Dissociation of functional roles of dynamin in receptor-mediated endocytosis and mitogenic transduction. *J. Biol. Chem.* 274, 24,575–24,578.

Whistler, JL, Chuang, H. H., Chu, P., Jan, L. Y. and von Zastrow, M. (1999) Functional dissociation of mu opioid receptor signaling and endocytosis: implication for the biology of opiate tolerance and addiction. *Neuron* 23, 737–746.

Whitaker-Azmitia PM, Borella A, Raio N. (1995) Serotonin depletion in the adult rat causes loss of the dendritic marker MAP-2. A new animal model of schizophrenia *Neuropsychopharmacology* 12, 269–272.

Wigge P, McMahon HT. (1998) The amphiphysin family of proteins and their endocytosis at the synapse. *TINS* 21, 339–344.

Wiklund NP, Cellek S, Leone AM, Iversen HH, Gustafsson LE, Brudin L, Furst VW, Flock Å, Moncada S. (1997) Visualization of nitric oxide released by nerve stimulation. *J. Neurosci. Res.* 47, 224–232.

Williams M, Jarvis MF. (2000) Purinergic and pyrimidinergic receptors as potential drug targets. *Biochem. Pharmacol.* 59, 1173–1185.

Williams RSB, Harwood AJ. (2000) Lithium therapy and signal transduction. *TIPS* 21, 61–64.

Williams SR, Stuart GJ. (2000) Action potential backpropagation and somato-dendritic distribution of ion channels in thalamocortical neurons. *J. Neurosci.* 20, 1307–1317.

Willmott NJ, Wong K, Strong AJ. (2000) A fundamental role for the nitric oxide-G-kinase signaling pathway in mediating intracellular Ca^{2+} waves in glia. *J. Neurosci.* 20, 1767–1779.

Wilson JX, Peters CE, Sitar SM, Daoust P, Gelb AW. (2000a) Glutamate stimulates ascorbate transport by astrocytes. *Brain Res.* 858, 61–66.

Wilson MT, Kisaalita WS, Keith CH. (2000b) Glutamate-induced changes in the pattern of hippocampal dendrite outgrowth: a role for calcium-dependent pathways and the microtubule cytoskeleton. *J. Neurobiol.* 43, 159–172.

Wong WT, Faulkner-Jones BE, Sanes JR, Wong ROL. (2000) Rapid dendritic remodelling in the developing retina: dependence on neurotransmission and reciprocal regulation by Rac and Rho. *J. Neurosci.* 20, 5024–5036.

Wood TL, O'Donnell SL, Levison SW. (1995) Cytokines regulate IGF binding proteins in the CNS. *Prog. Growth Factor Res.* 6, 181–187.

Woolley CS, Weiland NG, McEwen BS, Schwartzkroin PA, (1997) Estradiol increases the sensitivity of hippocampal CA1 pyramidal cells to NMDA receptor-mediated synaptic input: correlation with dendritic spine density. *J. Neurosci.* 17, 1848–1859.

Xia MQ, Hyman BT. (1999) Chemokines/chemokine receptors in the central nervous system and Alzheimer's disease. *J. Neurovirol.* 5, 32–41.

Yamagata K, Andreasson KI, Kaufmann WE, Barnes CA, Worley PF. (1993) Expression of a mitogen-inducible cyclooxygenase in brain neurons: regulation by synaptic activity and glucocorticoids. *Neuron* 11, 371–386.

Yan XX, Garey JJ. (1997) Morphological diversity of nitric oxide-synthesizing neurons in mammalian cerebral cortex. *J. Hirnforsch.* 38, 165–172.

Yang SN. (2000) Ceramide-induced sustained depression of synaptic currents mediated by ionotropic glutamate receptors in the hippocampus: an essential role of postsynaptic protein phosphatases. *Neuroscience* 96, 253–258.

Yao JK, Reddy RD, van Kammen DP, McElhinny LG, Korbanic CW. (1998a) Reduced level of the antioxidant proteins in schizophrenia. *Biol. Psychiat.* 42, 123S.

Yao JK, Reddy R, van Kammen DP. (1998b) Reduced level of plasma antioxidant uric acid in schizophrenia. *Psychiat. Res.* 80, 29–39.

Yao JK, Reddy R, McElhinny LG, van Kammen DP. (1998c) Reduced status of plasma total antioxidant capacity in schizophrenia. *Schiz. Res.* 32, 1–8.

Ye B, Lioa D, Zhang X, Zhang P, Dong H, Huganir RL. (2000) GRASP-1: a neuronal RasGEF associated with the AMPA receptor/GRIP complex. *Neuron* 26, 603–617.

Yee KT, Simon HH, Tessier-Lavigne M, O'Leary DM. (1999) Extension of long leading processes and neuronal migration in the mammalian brain directed by the chemoattractant netrin-1. *Neuron* 24, 607–622.

Yen G-C, Hsieh C-L. (1997) Antioxidant effects of dopamine and related compounds. *Biosci. Biotech. Biochem.* 61, 1646–1649.

Yermolaieva O, Brot N, Weissbach H, Heinemann SH, Hoshi T. (2000) Reactive oxygen species and nitric oxide mediate plasticity of neuronal calcium signaling. *PNAS USA* 97, 448–453.

Yoo AS, Cheng I, Chung S, Grenfell TZ, Lee H, Pack-Chung E, Handler M, Shen J, Xia W, Tesco G, Saunders AJ, Ding K, Frosch MP, Tanz RE, Kim T-W. (2000) Preseniin-mediated modulation of capacitative calcium entry. *Neuron* 27, 561–572.

Yoo BC, Kim SH, Cairns N, Fountoulakis N, Lubec G. (2001) Deranged expression of molecular chaperones in brains of patients with Alzheimer's disease. *Biochem. Biophys. Res. Com.* 280, 249–258.

Yoshizumi M, Abe J, Haendeler J, Huang Q, Berk BC. (2000) Src and Cas mediate JNK activation but not ERK1/2 and p38 kinases by reactive oxygen species. *J. Biol. Chem.* 275, 11,706–11,712.

Young CE, Arima K, Xie J, Hu L, Beach TG, Falkai P, Honer WG. (1998) SNAP-25 deficit and hippocampal connectivity in schizophrenia. *Cereb. Cortex.* 8, 261–268.

Young CE, Arima K, Falkai P, Honer WG. (2000) Synaptic proteins in the granule cell layer in schizophrenic hippocampus. *Schiz. Res.* 41, 105.

Yu, Z, Luo H, Fu W, Mattson MP. (1999) The endoplasmic reticulum stress-responsive protein GRP78 protects neurons against excitotoxicity and apoptosis: Suppression of oxidative stress and stabilization of calcium homeostasis. *Exp. Neurol.* 155, 302–314.

Yuste R, Majewska A, Holthoff K. (2000) From form to function: calcium compartmentalization in dendritic spines. *Nature Neuroscience* 3, 653–659.

Zacharias U, Nörenberg U, Rathjen FG. (1999) Functional intereactions of the immunoglobulin superfamily member F11 are differentially regulated by the extracellular matrix proteins tenascin-R and tenascin-C. *J. Biol. Chem.* 274, 24357–24365.

Zapf-Colby A, Olefsky JM. (1998) Nerve growth factor processing and trafficking events following Trk-A-mediated endocytosis. *Endocrinology* 139, 3232–3240.

Zhang B, Ganetsky B, Bellen HJ, Murthy VN. (1999) Tailoring uniform coats for synaptic vesicles during exocytosis. *Neuron* 23, 419–422.

Zhang L, Jope RS. (1999) Oxidative stress differentially modulates phosphorylation of ERK, p38 and CREB induced by NGF or EGF in PC12 cells. *Neurobiol. Aging.* 20, 271–278.

Zhao ZS, Khan S, O'Brien PJ. (1998) Catecholic iron complexes as cytoprotective superoxide scavengers against hypoxia: reoxygenation injury in isolated hepatocytes. *Biochem. Pharmacol.* 56, 825–830.

Zhou D, Lambert S, Malen PL, Carpenter S, Boland LM, Bennett V. (1998) AnkyrinG is required for clustering of voltage-gated Na channels at axon initial segments and for normal action potential firing. *J. Cell Biol.* 143, 1295–1304.

Zoccarato F, Cavallini L, Valente M, Alexandre A. (1999) Modulation of glutamate exocytosis by redox changes of superficial thiol groups in rat cerebrocortical synaptosomes. *Neurosci. Lett.* 274, 107–110.

Index